SpringerBriefs in Physics

Volume 1

CW00567914

For further volumes:
http://www.springer.com/series/8902

Eleftherios N. Economou

A Short Journey from Quarks to the Universe

Selected Solutions

Springer

Eleftherios N. Economou
University of Crete
FORTH IESL
N. Plastira Street 100
70013 Iraklion
Greece
e-mail: economou@admin.forth.gr

Additional material to this book can be downloaded from http://extra.springer.com

ISSN 2191-5423 e-ISSN 2191-5431
ISBN 978-3-642-20088-5 e-ISBN 978-3-642-20089-2
DOI 10.1007/978-3-642-20089-2
Springer Heidelberg Dordrecht London New York

Cover design: eStudio Calamar, Berlin/Figueres

Printed on acid-free paper

Springer is part of Springer Science+Business Media (www.springer.com)

There is, of course, an immense liberating role of Science at the central existential level. It is what Aristotle was saying about "θαυμάζειν". Science humanizes us, liberates us from our animal instincts, just because it makes us wondering and at the same time desiring to explain… Yet it shows us our limits and our mortality… Thus Science is something immeasurably precious…. Science can help us approach anew the real poetic and mythical dimension of human existence.

C. Castoriadis

To Christos, Eleni, Zinovia, Yiannis, and Evropi with wishes "για καλή στράτα" in life's pathway.

Suggested by A., of course

Preface

This short book grew out of lectures presented to different audiences (physics students, physicists, material scientists, engineers) and on various occasions (colloquia and seminars in physics and other departments, conferences, special events). The main purpose of these lectures and, obviously, of the present book is to show that basic formulae concerning the various structures of the *physical world* pop out quickly, if some *basic ideas*, the *universal physical constants, and dimensional considerations* are exploited. Of course, as R. Feynman pointed out, "a little thinking has to be applied too".

The basic ideas include the three cornerstones of science, namely the atomic idea, the wave-particle duality, and the minimization of free energy as the necessary and sufficient condition for equilibrium (these are presented in Chaps. 2, 3, and 4 respectively). These fundamental ideas exhibit their worth when accompanied by the values of the physical constants: the universal ones, $\hbar$, c, the coupling constants of the four interactions, G, e, g_w, g_s and the masses of the elementary particles, m_p, m_n, m_e, m_w,... An important consequence of the atomic idea is that the relevant (for each case) physical constants will appear in the quantities characterizing the various structures of the world (microscopic or *macroscopic*). Combining this last observation—often overlooked—with dimensional analysis, presented in Chap. 5, and "a little thinking", one can obtain, in several cases, an amazing short-cut derivation of formulae concerning the various structures of nature from the smallest (baryons and mesons) to the whole Universe, as shown in Chaps. 6–13. In each one of these 8 chapters, in parallel with a demonstration of the method just outlined, a *condensed* (sometimes too condensed) introduction to the relevant subject matter together with a few physical remarks are presented.

I must admit that the main fronts on which our horizons are widened, namely the *small*, the *large*, and the *complex* could not be treated even remotely adequately in this short book. Actually the *complex*, as represented by the living matter, was too complex for our simple method; so it was left out completely (however, see the epilogue). The *large* (cosmology) and the *small* (elementary particles) tend to converge to a unified subject (the snake in Fig. 1.1, p. 2, is biting its tail) fed with novel observational data from special instruments mounted usually on satellites, and boosted by high experimental expectations from the Large Hadron Collider. Nevertheless, in these fields there are several open fundamental questions concerning conditions well beyond our present or near future experimental capabilities. This vacuum of confirmed knowledge is filled with new, intriguing, imaginative ideas and novel proposed theories (such as supersymmetry, string theory, M-theory, see reference [P1]) which, if established, will radically change our world view. In spite of the wider interest in these ideas and theories and their high intellectual value, I decided for several reasons to restrict myself in the present book to experimentally or observationally tested ideas and theories.

The intended readers of this book are senior undergraduate or graduate students in Physics, Engineering, Applied Mathematics, Chemistry, and Material Science. They may find the book a useful supplement to their courses as a concise overall picture of the physical world. Research physicists, physics teachers, and other scientists may also find this short book intellectually stimulating and entertaining. The required background is no more than a *working* familiarity with the first year Science or Engineering material.

I am deeply indebted to my colleague, Prof. V. Charmandaris, for his encouragement during the writing of this book and for reading my entire manuscript and making many useful suggestions. Of course, whatever misprints or misrepresentations remained are my own responsibility only. I am also grateful to Ms Maria Dimitriadi for her invaluable help in bringing my manuscript to its final form.

January 2011 Heraklio

Contents

Part II Structures Mediated by Strong Interactions

Part III The Realm of Electromagnetism

Part IV Gravity at Front Stage

Chapter 1
Introduction: The World According to Physics

Ah, but a man's reach should exceed his grasp. Or what's a heaven for?

R. Browning

Abstract In this introductory chapter the subject matter as well as the methodology of Physics are briefly presented together with a list of the basic structures of matter.

Keywords Natural Philosophy · Structures of matter · Origin of matter · Nature of Physics

Modern-era Physics started as Natural Philosophy. As this former name implies, Physics is built (and continues to be developed) around the age-old, yet ever-present questions:

- What the World and its parts are made of? How?
- Is there a hidden underlying simplicity in its immense complexity and diversity?

The first question implies that the subject matter of Physics is the World, both natural and man-made; from its smallest constituent to the whole Universe (see Fig. 1.1 and Table 1.1 p. 3). In this sense, Physics is encompassing the other natural sciences (such as Chemistry and Biology) and even Engineering, while at the same time it provides also their foundation. What allows Physics to have this dual role is its characteristic methodology. The latter is precise and quantitative, yet capable of abstraction (therefore mathematical). It is based on observations and well controlled experiments both as sources of ideas and as tests for falsification or tentative confirmations of proposed and—even—established theories. Moreover, as the second question suggests, the methodology of physics requires the formulation of a few fundamental quantitative relations on which everything else is based. These features of the methodology of Physics account for its role as foundation of every other science and engineering, but also for its limited

E. N. Economou, *A Short Journey from Quarks to the Universe*,
SpringerBriefs in Physics, 1, DOI: 10.1007/978-3-642-20089-2_1,

penetration into very complex, yet very important, parts of the World (such as the molecular and the biological worlds). This leaves plenty of space to more specialized sciences such as Chemistry and Biology and, of course, Engineering.

Over the last 50 years or so Physics is actively concerned over another fundamental, age-old, but much more difficult question which stretches its methodology to the limit:

- How did the World start, how did it evolve, and where is it going?

Detailed observational data, such as the recession of distant galaxies at a speed proportional to their distance from Earth, the spectral and angular distribution of the Cosmic Background Microwave Radiation, etc., combined with established physical theories, allowed us to reconstruct roughly some of the main events in the history of the Universe. Naturally, other crucial events, including the emergence of life, remain unknown and they are the subject of current research. Subject of current theoretical research is also the development of a successful quantum theory of gravity, which is expected to let us approach the very moment of the genesis of the Universe.

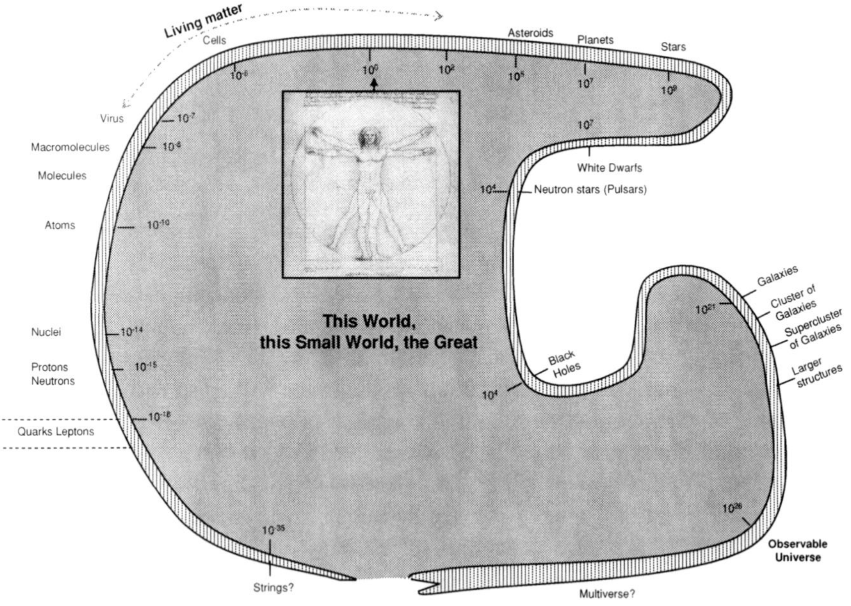

Fig. 1.1 The main structures of matter from the smallest to the largest size (*clockwise*) and the suspected connection of the two extremes (see Ref. [1.2]). The indicated sizes are in meters

Table 1.1 Levels of the structure of matter (see also [1.1])

Level of the structure of matter	Length scale (in m)	Constituents	Interaction(s) responsible for the structure
Quarks	$<10^{-18}$	It seems to be elementary	–
Electron	$<10^{-18}$	It seems to be elementary	–
Proton	10^{-15}	u,u,d quarks	Strong, weak,, E/M
Neutron	10^{-15}	u,d,d quarks	Strong, weak,, E/M
Nuclei	10^{-15}–10^{-14}	Protons, neutrons	Strong, E/M, weak,
Atoms	10^{-10}	Nucleus, electrons	E/M
Molecules	$>10^{-10}$	Atoms and/or ions and electrons	E/M
Solids (primitive cell)	$>10^{-10}$	Atoms and/or ions and electrons	E/M
Cells	$\geq 10^{-6}$	Molecules	E/M
Biological entities (e.g., Homo Sapiens)	10^{-8}–10^{2} (10^{0})	Molecules, cells, tissues, organs, microbes	E/M
Planets	10^{6}–10^{7}	Solids, liquids, gases	E/M, Gravitational
Stars	10^{9}	Electrons, nuclei, ions, photons	Gravitational, strong, weak, E/M
White dwarfs	10^{7}	Nuclei, electrons	Gravitational
Neutron stars	10^{4}	Neutrons and some protons and electrons	Gravitational
Astrophysical black holes	10^{4}	?	Gravitational
Galaxies	10^{21}	Stars, ordinary and dark matter, photons, neutrinos	Gravitational
Observable universe	10^{26}	Galaxies, dark matter, dark energy	Gravitational, others?

In concluding these introductory remarks, we present below some of the various branches of Physics and their correspondence and/or overlap with more specialized sciences as well as some examples of the impact of Physics on important technologies:

Elementary Particle Physics	
Nuclear Physics	Chemistry
Atomic and Molecular Physics	
Condensed Matter Physics	Material Science
Biophysics	Biology
Geophysics	Geology
Atmospheric and Space Physics	Meteorology, Global Climate
Astrophysics	Astronomy
Cosmology	
E/M Waves	Telecommunications
Lasers	
Solid State devices	
Integrated circuits	Computers
Magnetic devices	
X-rays	
γ-rays	Medical Technologies
Magnetic resonance (MRI)	
Positron annihilation (PET)	

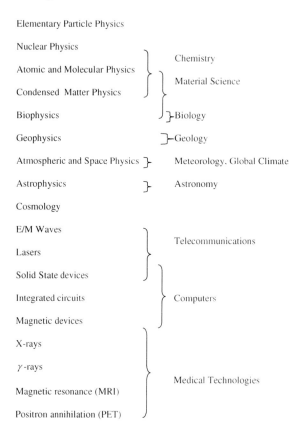

Part I
Three Key-Ideas and a Short-Cut

Chapter 2
The Atomic Idea

If, in some cataclysm, all of scientific knowledge were to be destroyed, and only one sentence passed on to the next generation of creatures, what a statement would contain the most information in the fewest words? I believe it is the atomic hypothesis that all things are made of atoms–little particles that move around in perpetual motion, attracting each other when they are a little distance apart, but repelling upon being squeezed into one another. In that one sentence, there is an enormous amount of information about the world, if just a little imagination and thinking are applied.

R.P. Feynman, *The Feynman Lectures on Physics.*

Abstract In this chapter the elementary particles from which all things are made are presented together with the interactions which bring them together. The latter are transmitted by indivisible particle-like entities. The so-called Feynman diagrams describing the various interactions are introduced.

Keywords Electron · Leptons · Quarks · Spin · Hadrons · Interactions

Summary and a Comment According to the atomic idea everything is made of indestructible, indivisible elementary particles (to be called here *m-particles*[1]) which attract each other to form composite structures. To substantiate the atomic idea we must answer the following questions:

a. What are the various kinds of *m*-particles and which are their properties?
b. What kind of interactions do lead to their attraction[2]?
c. What does counterbalance this attraction and establishes equilibrium?

In this chapter we provide some answers to questions (a) and (b); (c) will be examined in the next chapter.

The atomic idea is the decisive step towards the sought-after underlying simplicity in the World; this is so, because the immerse complexity and diversity of the World can in principle be deduced from the *properties* and the *microscopic*

---

[1] *m* stands for matter.

[2] It turns out that interactions are actually transmitted through indivisible elementary quantities to be named here (for distinguishing them from the *m*-particles) *interaction-carrier-particles* (ic-particles). The ic-particles, in addition to mediating forces, are capable of making certain transformations of the *m*-particles among each other; thus the indestructibility of elementary *m*- or ic-particles is of questionable validity.

E. N. Economou, *A Short Journey from Quarks to the Universe*,
SpringerBriefs in Physics, 1, DOI: 10.1007/978-3-642-20089-2_2,

motion, under the influence of interactions, of a few kinds of elementary *m*-particles presented in Table 2.1. To state it differently, the atomic idea implies that properties of macroscopic matter ought to be expressed, at least in principle, in terms of a few numbers characterizing the elementary particles and their interactions. This is true as long as the considered properties of macroscopic matter do not depend on some uncontrolled initial conditions.

2.1 The Elementary Particles of Matter

As shown in Table 2.1, all *m*-particles have spin[3] 1/2. This is an important characteristic property which distinguishes the *m*-particles from the ic-particles; the latter have integer spin, as shown in Table 2.2 (p. 10). In Table 2.1 the rest mass is multiplied by c^2 to be given in energy units (1 MeV $= 10^6$ eV $= 1.6022 \times 10^{-13}$ J). The rest mass of each neutrino is not zero, but its value has not been determined (the numbers given in Table 2.1 are arbitrary guesses consistent with established upper limit). It is useful to introduce the concept of lepton and baryon number: All leptons have lepton number 1 and baryon number 0: all quarks have lepton number 0 and baryon number 1/3 so that protons, neutrons and other composite entities consisting of three quarks[4] to have baryon number 1. The charge e of proton is taken as the unit of charge ($e = 1.6021765 \times 10^{-19}$ C).

Quarks have never been observed as isolated free particles; they have been seen always as constituents of composite entities such as baryons (e.g., protons or neutrons) or *mesons.*[5] This is the reason for which the rest mass of the quarks, especially the lighter ones, is not known accurately.

We should point out that for each *m*-particle of Table 2.1 there is an *m*-antiparticle which is denoted with a bar above the symbol of the corresponding particle. So $\bar{\nu}_e$ is the antiparticle of the e-neutrino and $\bar{d}$ is the antiparticle of the down quark. For historical reasons, the antiparticle of electron is denoted as e^+ and it is called *positron*. Each antiparticle has exactly the same rest mass and spin but exactly the opposite electric charge, the opposite lepton number, and the opposite baryon number of the corresponding particle. Note that antiparticles do not participate in the structure of matter, since a pair of same particle/antiparticle sharing the same volume is annihilated by giving rise to photons[6] and/or pair of neutrinos/

---

[3] Spin times the universal constant $\hbar$ ($\hbar = 1.0545716 \times 10^{-34}$ J · s) gives the intrinsic angular momentum of each *m*-particle or ic-particle.

[4] Composite entities consisting of three quarks are called *baryons*.

[5] *Mesons* are composite entities consisting of one quark and one antiquark. Baryons and mesons are collectively called *hadrons*.

[6] The medical diagnostic method known as Positron Emission Tomography (PET) is based on the introduction of positrons which, subsequently, by meeting electrons annihilate giving rise to two photons of opposite direction and the same energy each (equal to the rest mass of electrons, 0.51 MeV).

Table 2.1 The established elementary *m*-particles are of two types (leptons and quarks) and of three families (1st, 2nd, 3rd) [2.1].

Leptons					Quarks				
	Symbol, name	Mass$\times c^2$ (MeV)	Electric charge	Spin	Symbol, name	Mass$\times c^2$ (MeV)	Electric charge	Spin	Baryon number
1st	ν_e, e-neutrino	$\sim 10^{-9}$?	0	1/2	*u*, up quark	1.5-4.5	2/3	1/2	1/3
	e^-, electron	0.51	−1	1/2	*d*, down quark	5-8.5	−1/3	1/2	1/3
2nd	ν_μ, μ-neutrino	$\sim 10^{-8}$?	0	1/2	*c*, charm quark	1,000–1,400	2/3	1/2	1/3
	μ, muon	105.7	-1	1/2	*s*, strange quark	80–155	−1/3	1/2	1/3
3rd	ν_τ, τ-neutrino	$\sim 10^{-7}$?	0	1/2	*t*, top quark	174,000	2/3	1/2	1/3
	τ, tau	1,777	−1	1/2	*b*, bottom quark	4,000–4,500	−1/3	1/2	1/3

Only three of the *m*-elementary particles (e^-, *u*, *d*) are participating in the structure of ordinary matter. Neutrinos are too light and weakly interacting to be bound in ordinary matter. Particles of the 2nd and 3rd family, having excessive rest energy, are metastable and decay to particles of the 1st family. More exotic particles, which have not been observed yet, may account for the dark matter and, possibly, for the dark energy; the existence of the latter is supported by astrophysical and cosmological observations. (see also Ref. [2.1])

Table 2.2 The four basic interactions and their interaction-carrier particles [2.1]

Name	Dimensionless strength	Range (m)	Type	ic-particle name-symbol	Rest mass (MeV)	e-charge	Spin	Emitters/Absorbers
Gravitational	$a_G \equiv \frac{G\,m_p^2}{\hbar\,c} = 5.9 \times 10^{-39}$	∞	Attractive	Graviton?	0	0	2	All particles
Electromagnetic	$a \equiv \frac{e^2}{\hbar\,c} = \frac{1}{137}$	∞[a]	Attractive or repulsive	Photon, γ	0	0	1	electrically-charged particles
Weak "nuclear"	$a_w \equiv \frac{g_w^2\,\sqrt{2}\,m_p^2}{\hbar\,c\ \ m_w^2} \approx 10^{-5}$	10^{-18}	$u \rightleftarrows d + \ldots$	Vector bosons Z^o	91,000	0	1	All m-particles photons, vector bosons
				Vector bosons W^+	80,000	1	1	
				Vector bosons W^-	80,000	−1	1	
Strong "nuclear"	$a_s \equiv \frac{g_s^2}{\hbar\,c} \approx 1$	10^{-15}	Attractive or repulsive	Gluons, g	0	0	1	Quarks and gluons

The physical symbols appearing in the second column of this table are as follows: G is the gravitational constant, m_p is the mass of the proton, $\hbar$ is Planck's constant, "trade mark" of Quantum Mechanics, c is the velocity of light in vacuum, g_w is the strength of the weak interaction, m_w is the mass of the W^+ vector boson, and g_s is the strength of the strong interaction. Numerical values of some of these quantities are given in Table 1 in p. 141. The range r_o of each interaction is connected to the rest mass m of the corresponding ic-particle through the formula $r_o = \hbar/m\,c$. The strong interactions to be discussed in Chap. 6, are exceptions

[a] Because the E/M interaction can be either attractive or repulsive the residual E/M interaction between neutral bodies becomes short range of the order of a few Angstroms

antineutrinos. However, antiparticles are produced artificially and their properties have been determined experimentally in spite of their short life-time.

It must also be pointed out that quarks besides their electric charge, carry another type of charge which will be called here *color-charge*, or c-charge, although there is no relation with the usual color). The c-charge is the source and the acceptor of the strong force, the same way that the electric charge is the source and the acceptor of the electromagnetic (E/M) force (see Table 2.2, p. 10). In contrast to the electric charge, which is of one type, the c-charge is of three different types denoted as *red* (R), *green* (G), and *blue* (B). The opposite charges of these three types of c-charges are called *antired*, *antigreen*, and *antiblue* and are denoted by $\bar{R}$, $\bar{G}$, $\bar{B}$, respectively. Each quark carries an R, or a G, or a B, c-charge. Each antiquark carries an $\bar{R}$, or $\bar{G}$, or $\bar{B}$, c-anticharge. The combination of a pair of quark/antiquark of the form $R\,\bar{R}$, or $G\bar{G}$, or $B\bar{B}$ has zero c-charge (in other words, it is colorless). Similarly colorless is the combination of three quarks of c-charges R, G, and B, respectively, or of three antiquarks of c-anticharges $\bar{R}, \bar{G}, \bar{B}$ respectively. In view of this property the notation R, G, and B for the three types of c-charge is not unreasonable, since it is reminiscent of Newton's disc, where the equal combination of red, green, and blue colors produces white. Following this analogy with real colors, we could say that an antiblue c-charge, which is equivalent to the RG combination of c-charges, could be presented as a "yellow" c-charge [2.1].

We should stress once more that out of the 12 elementary m-particles of Table 2.1 and the corresponding twelve m-antiparticles, only three (the electron, the up quark, and the down quark) participate in the structure of all ordinary matter surrounding us and in us.

2.2 The Interactions and Their Elementary Carrier-Particles

There are four basic established interactions: The gravitational, the electromagnetic, the weak (nuclear), and the strong (nuclear). All forces among m-particles are mediated through interactions as follows: An m-particle emits an ic-particle which in turn is absorbed by another m-particle; this causes a force between the two m-particle. Sometimes the emission or the absorption of an ic-particle may transform the m-particle to another m-particle. An ic-particle besides disappearing by being absorbed, may disappear by creating a pair of m-particle/m-antiparticle. All these elementary processes are subject to certain conservation laws, meaning that certain quantities must be conserved, i.e., must be the same before and after each of these elementary processes. The electric charge is such a quantity, so that the sum of the electric charges of the particles entering an elementary process must be the same as the sum of the electric charges of the particles coming out of this process. Other conserved quantities, such as the electric charge, are the momentum or the angular momentum, the color-charge, the baryon number, the lepton number (separately for each of the three families[7]). Energy too must be conserved *overall*, meaning that the total energy of the initial particles entering a *physical* process must be the same as the energy of the final particles, after this process is over. Some interactions, such as the strong one, conserve additional quantities as well. The conservation laws impose severe restrictions regarding which physical processes are allowable and which are forbidden.

In Table 2.2 we present some properties of the four interactions and of their elementary ic-particles

There are several points to be noticed in Table 2.2: The strength of the gravitational interaction is unimaginably weaker than any other interaction. To stress this point take into account that when you lift an object, such as a chair, the electric force generated by your muscles overcomes the gravitational force exercised on the chair by the whole Earth.

The extreme weakness of gravity compared to the other three forces can be compensated in systems, such as stars and planets, consisting of a very large number of *m*-particles as a result of its long range and always attractive character. Neither the gravitational waves nor their ic-particles, the gravitons, have been detected experimentally, because they carry such a minute amount of energy. Finally, keep in mind that all particles, both the *m*- and the ic-ones are emitters and absorbers of the gravitational force, since all of them possess energy and, hence, relativistic mass.

The E/M interaction has some unique features: First, it dominates the structure of matter from the scale of atoms all the way to that of an asteroid, i.e., it reigns over 15 orders of magnitude in linear dimensions. Second, its ic-particle, the photon, is the only one that, besides mediating a force, can also travel free in space carrying over long distances both energy and information. No other ic-particle can do that: Gravitons are too light to be noticed; vector bosons decay too fast; and gluons are trapped forever within baryons, mesons, or any other composite quark structure. Third, photons can easily be emitted and detected. These unique features account for the dominant role of photons in the World, including their biological and technological functions. Photons deserve further examination; so we shall return to them in Chap. 5 (pp. 33–38). For the time being, we present in a diagrammatic way their role as mediators of the E/M force (in Fig. 2.1) and their function as free particles in various processes of physical and/or technological importance (Fig. 2.2).

The main function of the weak interaction is to transform a quark or a lepton to another quark or to another lepton, respectively, by the emission of one of its corresponding three ic-particles which collectively are called *vector bosons* (Z°, W^{+}, W^{-}). Almost immediately the emitted ic-particle decays to a pair of quark/antiquark or to a pair of lepton/antilepton. Thus all *m*-particles (or their antiparticles) are subject to the weak interaction as well as the photons and all the vector

[7] There are some rare processes which violate this last statement; e.g. an e-neutrino may change to an μ-neutrino and vice versa.

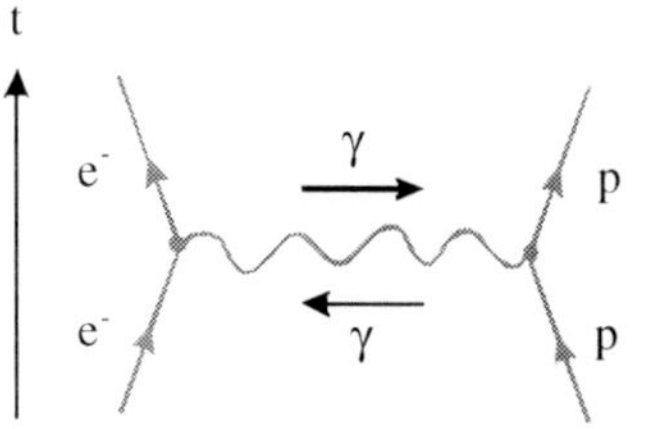

Fig. 2.1 The exchange of a photon between two electrically charged particles creates a force between them. In this case the electron emits a photon and the proton absorbs it, or vice versa. This force is associated with the so-called potential energy V, which in the present case is given approximately by the well known *Coulomb formula*: $V = -e^2/r$, where e is the electric charge of the proton, $-e$ is the charge of the electron and r is their distance. For $r = 0.53 \times 10^{-10}$ m, $V = 27.2$ eV. The photon in this case does not appear in either the initial or the final state; when this happens, the particle is called a *virtual* one (here a virtual photon). Throughout this book the so-called Gauss-CGS (G-CGS) system is used for E/M quantities, because it is physically more transparent, since in it the velocity of light appears explicitly in Maxwell's equations. On the other hand, the widely used system called SI employs more convenient and familiar units. The correspondence of some quantities appearing in the SI with those appearing in G-CGS is as follows (first the SI and second the G-CGS): $\varepsilon_o \leftrightarrow 1/4\pi$; $E \leftrightarrow E$; $D \leftrightarrow D/4\pi$; $\mu_o \leftrightarrow 4\pi/c^2$; $B \leftrightarrow B/c$; $H \leftrightarrow cH/4\pi$. To transform a formula involving $e^2$from G-CGS to SI replace e^2 *by* $e^2/4\pi\varepsilon_o$

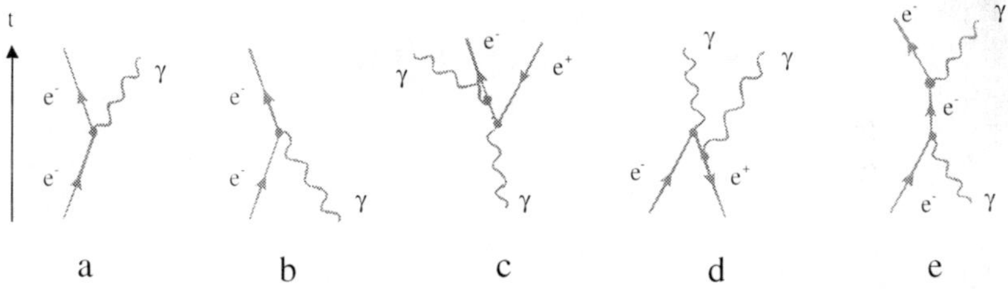

Fig. 2.2 Processes in which a free photon is participating: **a** Emission of a photon by a deaccelerating electron, as in X-rays production, **b** absorption of a photon by an electron, as in the photoelectric phenomenon, **c** creation of an electron/positron pair by absorbing an incoming photon and emitting another photon, **d** annihilation of an electron/positron pair with the emission of a pair of two photons, as in the diagnostic method PET, **e** absorption of a photon and emission of another photon from an electron, as in Raman scattering

bosons. Notice how massive the latter are, a fact which explains the extreme short range of the weak interaction. In Fig. 2.3 a d quark is transformed into a u quark by the emission of a W^- vector boson which in turn decays into a pair of electron/e-antineutrino. The whole process, known as beta-decay, is changing a neutron (u,d.d) to a proton (u,u,d) with the emission of an electron and an e-antineutrino. This beta decay is taking place in the byproducts of a nuclear reactor and is responsible for their radioactivity. In most cases the beta-decay is followed by the emission of a very energetic photon, an event known as gamma radioactivity.

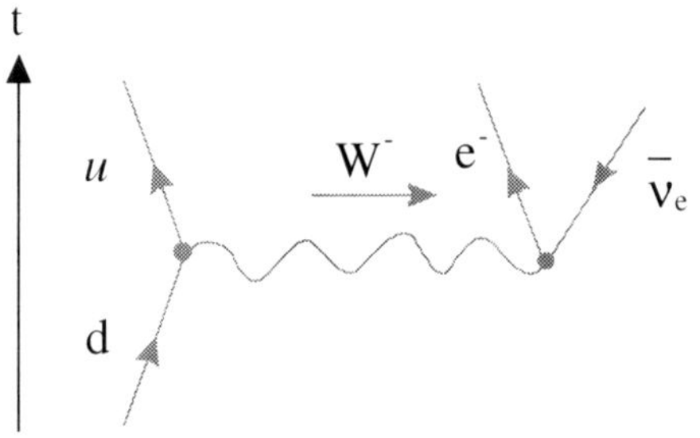

Fig. 2.3 The elementary process leading to the so-called β-decay. The weak interaction is responsible for this process, as the emission and subsequent absorption of a W^- vector boson implies. Note that antiparticles, as the e-antineutrino in the present case, are denoted in the diagrams with an arrow of opposite direction to the arrow of time. Diagrams, as those of Figs. 2.1, 2.2, 2.3. etc. are known as Feynman diagrams; they allow, through some explicit but rather complicated rules quantitative calculations such as the life-time of a free neutron

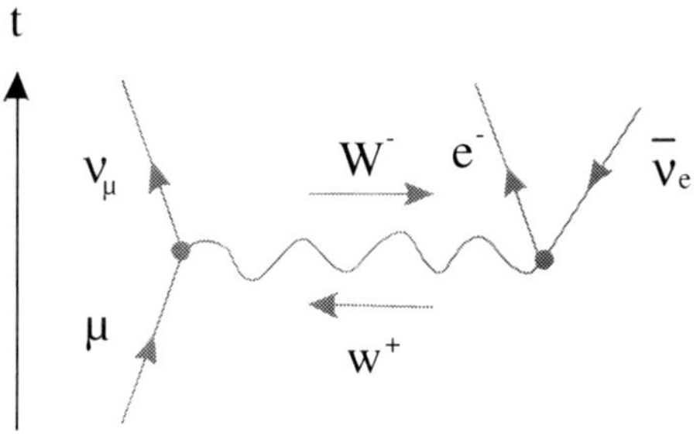

Fig. 2.4 A muon (μ) is transformed to μ-neutrino (ν_μ) by emitting a W^-vector boson. Subsequently, the W^- decays to an electron/e-antineutrino pair. This is how the muon decays

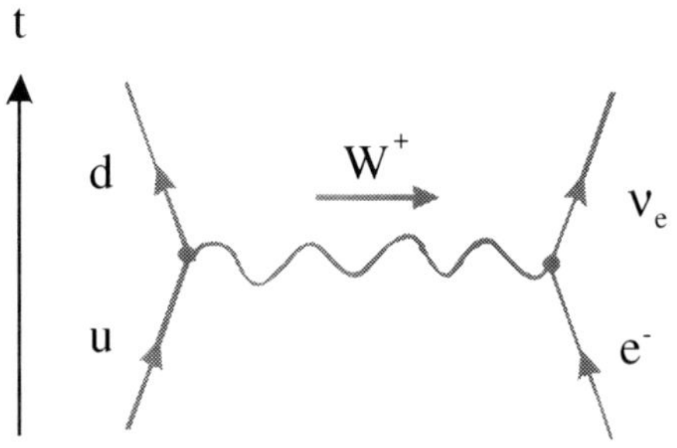

Fig. 2.5 A u quark is transformed into a d quark by emitting a W^+vector boson, which in turn meets a preexisting electron and the pair W^+/electron disappears giving rise to an e-neutrino (ν_e)

In Fig. 2.4 the process of a muon decay to an electron, an e-antineutrino, and a μ-neutrino taking place through the weak interaction is shown. In Fig. 2.5 the process of a u quark and an electron being transformed to a d quark and an e-neutrino through the exchange of a charged vector boson is shown. This process takes place in some heavy nuclei changing a proton to neutron with the absorption of an inner orbiting electron and the emission of an e-neutrino.

In all processes, and at each vertex where a particle-carrier of any interaction is emitted, or absorbed, or decays, etc., certain quantities, such as the momentum or the angular momentum, the electric charge, the color charge, the baryon number, the lepton number must be conserved (for each family separately). This means, e.g., that for each vertex, the total baryon number in must be equal to the total baryon number out. Notice that the energy does not need to be conserved at each elementary vertex. However, it must be conserved overall, meaning that the total energy before any reaction starts must be the same as the total energy after this reaction is concluded.

We are coming finally to the strong interactions. The corresponding ic-particles are the gluons, which bind together three quarks to form the so-called baryons, such as the proton (consisting of two u quarks and one d quark, all of different color-charge as to make the proton colorless) or the neutron (consisting of two d quarks and one u quark, all of different color-charge as to make the neutron colorless). Free protons, in contrast to free neutrons, are stable. Neutrons are also stable within non-radioactive nuclei. Of course, there are many other short-lived baryons, all of them colorless. A quark and an antiquark through the exchange of gluons form short-lived combinations called mesons. Mesons also are colorless. Mesons consisting of the first family of quarks and antiquarks are called pions. There are three types of pions: $\pi^+ \equiv u\bar{d}$, $\pi^- \equiv d\bar{u}$, and $\pi^o \equiv (u\bar{u} - d\bar{d})/\sqrt{2}$. There are eight types of gluons of different combinations of c-charge/c-anticharge as follows: $R\bar{G}$, $R\bar{B}$, $G\bar{R}$, $G\bar{B}$, $B\bar{R}$, $B\bar{G}$, and two more $(R\bar{R} - G\bar{G})/\sqrt{2}$, $(R\bar{R} + G\bar{G} - 2B\bar{B})/\sqrt{6}$.

We mention here that the accepted theory of elementary particles, called the *standard model*, requires the existence of another ic-particle, known as the *Higgs particle*. Virtual Higgs particles interacting with the various elementary particles give rise to the mass of all particles. The Higgs particle has not been detected as of this writing (January, 2011). According to the standard model, its rest energy is expected to be in the range between 114,000 and 158,000 MeV and its spin and its electric charge must be zero. Because of its very large rest mass the creation of Higgs requires very high energies. The new accelerator at CERN is expected to reach such energies and, hence, to produce and detect the Higgs particle, providing thus another confirmation to the standard model.

In spite of the fact that the standard model accounts for all data concerning reactions among elementary and/or composite particle, there are deeper questions which remain unanswered: Why there are three and only three families of elementary *m*-particles? Why do the rest masses of elementary particles have the values they have? Why do the dimensionless strengths of the interactions have the values they have? Is there a deeper level at a much smaller scale where such questions may find satisfactory answers? There is no lack of ideas and proposed theoretical schemes going beyond the standard model designed to provide a more unified world-view even at the expense of introducing more abstract concepts and less familiar mathematics. One such scheme goes under the name of supersymmetry [1.1] and predicts the existence of more elementary particles: one additional fermion for each boson of the standard model and one additional boson for each fermion. It predicts also an energy of unification of the forces in the sense that the coupling constants of all interactions have the same value at this energy. Another scheme known as string theory [1.1] abandons the idea that the ultimate elementary building units of the World are point-like particles; instead it introduces extended objects such as strings and/or membranes at a scale of 10^{-35} m, i.e., 17 orders of magnitude smaller than our present day capabilities. This scheme has several attractive features as to be actively studied over the last 30 years or so, in spite of the lack of experimental evidence supporting it [2.1].

Returning now to the established world-view of the hierarchical structure of matter, we can summarize the situation as follows: By combining three quarks through virtual gluons, we obtain two composite *m*-particles, the proton and the neutron, as a first step in the long journey towards the enormous variety of composite stable formations of the World. Following this first step, we continue with protons and neutrons attracting each other through residual strong interactions to form the various nuclei. The free protons and all the other composite nuclei, having positive electric charge attract the negatively charged electrons and form electrically neutral atoms. As we shall see, atoms attract each other through residual electrical interactions to form an enormous number of different molecules. This process continues until astronomical objects are formed, where all their matter is held together by the always attractive gravitational force. The crucial question is why all these attractive interactions do not continue to squeeze the various systems without limit, the same way that gravity squeezes the big dead stars to black holes? What does it change the attraction to repulsion upon the particles "*being squeezed into one another*" to quote Feynman? What does it oppose the attractive interactions so that physical systems end up in stable equilibrium states? We thus returned to where we started, namely to the third question posed in the summary of this chapter:

2.2.1 What Does Counterbalance the Overall Attraction of the Interactions and Establishes Equilibrium?

Feynman implicitly suggested the answer to the above question by mentioning that atoms are "*little particles that move around in perpetual motion*". Indeed, perpetual motion of the "*little particles*" implies a pressure which tends to blow the system apart. If this pressure becomes equal to the squeezing pressure of the attractive forces, the system may come to equilibrium. By following this line of thought, it is only natural to ask ourselves questions such as these: How this perpetual motion comes into being? What physical mechanism produces it? The answers are found in the most fundamental, yet difficult to accept and digest, empirical discovery, that of *the wave- particle duality*, to be briefly presented in the next chapter.

Chapter 3
The Wave-Particle Duality

So reasonable the incomprehensible.

O. Elytis

Abstract The wave-particle duality asserting that everything possesses partly wave and partly particle properties is presented. This duality, on which Quantum Mechanics is based, provides the required kinetic energy to counterbalance the attractive nature of the interactions and stabilizes both the microscopic and the *macroscopic* structures of matter. The core of Quantum Mechanics can be condensed to the following three fundamental principles possessing amazing quantitative predictive power: Those of Heisenberg, Pauli, and Schrödinger.

Keywords Waveparticle · QM · Uncertainty · Exclusion · Perpetual motion · Discreteness

Summary and Comments According to the wave-particle duality, a particle of energy ε and momentum $\boldsymbol{p}$ does not follow a trajectory, as classical physics asserts, but propagates as a wave of angular frequency

$$\boxed{\omega = \varepsilon/\hbar} \tag{3.1}$$

and of wavevector[1]

$$\boxed{\boldsymbol{k} = \boldsymbol{p}/\hbar}. \tag{3.2}$$

where $\hbar$ is one of the most fundamental physical constants called Planck's constant. Thus, this "particle" is not a classical one, but what it could be called a waveparticle. Moreover, a wave of angular frequency ω and wavevector $\boldsymbol{k}$ consists of discrete, indivisible entities of energy $\varepsilon = \hbar\omega$ and momentum $\boldsymbol{p} = \hbar\boldsymbol{k}$. It is only fair to call such a "wave" particlewave. Quantum Mechanics synthesizes the apparently contradictory concepts of *particle* and *wave* to a consistent, yet at odds

---

[1] The magnitude of k is related to the wavelength λ through the relation, $k = 2\pi/\lambda$, and its direction gives the direction of propagation of the wave.

E. N. Economou, *A Short Journey from Quarks to the Universe*,
SpringerBriefs in Physics, 1, DOI: 10.1007/978-3-642-20089-2_3,

with our perceptions, theory which can be distilled into three fundamental principles: *uncertainty, exclusion, spectral discreteness.* We show that the first two of these principles account quantitatively for the required perpetual microscopic motions necessary to counterbalance the squeezing pressure of the attractive forces and establishe, thus, equilibrium. The third principle makes composite microscopic structures to behave like elementary ones up to a limit. This keeps an admirable balance between the need for stability and the need for change.

The wave-particle duality, in spite of the initial strong resistance to its counterintuitive features, has been established by numerous hard facts as one of the more basic laws of Nature. This duality asserts that the actual entities of the World are endowed with some of the particle properties, such as discreteness and, hence, indivisibility when elementary, *as well as* some of the properties characteristic of waves, such as non-locality which is a necessary condition for the phenomenon of interference. Thus, a more appropriate name—to be used in this book—for these entities is waveparticles or, equivalently, particlewaves. The essence of the wave-particle duality is captured in Feynman's description of the *gedanken* two slit experiment [3.1]. This idealized version of a real experiment shows beyond any doubt that the familiar macroscopic laws of motion are completely inapplicable to the microscopic motions of microscopic particles. Moreover, the experiments clearly show that waveparticles exhibit the characteristic wave phenomenon of interference and thus lead to the wave-particle duality. The implications of the latter, and, hence, of Quantum Mechanics, are not restricted only to the microscopic world, but they extend to all structures of the World from the smallest to the largest. This follows directly from the atomic idea which attributes the properties of any part of the World, macroscopic or not, to the properties and the microscopic motions of the microscopic constituents of matter. We would like to stress this point emphatically:

3.1 The Properties of the World at Every Scale are of Quantum Nature: If they were Not, We Would Not Exist

The validity of this statement becomes apparent by noticing[2] that the observed values of macroscopic properties, such as density, compressibility, specific heat, electrical resistance of solids and liquids as well as the mass of the stars, and their final fate (whether white dwarfs, neutron stars, black holes) involve the universal constant $\hbar$, the "trade mark" of particle-wave duality and of Quantum Mechanics (QM). If $\hbar$ were zero, as in the classical worldview, the present structures of the World would collapse to the classical analogue of a black hole. Of course there are some *isolated* cases where Classical Mechanics (CM) produces results in very good agreement with the quantum mechanical ones. The specific heat of insulators at elevated temperatures and the *macroscopic motions*, such as sea waves or planets

[2] See Chaps. 6–13.

orbiting the Sun, are more than adequately accounted for by CM, without this meaning that QM fails in these cases. Simply, in these cases QM reduces to CM. We must also clarify that there are some important aspects of QM [3.2–3.5], such as coherence, a prerequisite for wave interference, and the peculiar correlations present in the so-called entangled systems (see, e.g., the Einstein–Podolsky–Rosen gedanken experiment [3.6] and its realization by Aspect et al. [3.7]), which are extremely difficult to maintain in macroscopic systems. However, this by no means implies that the macroscopic world is classical. We repeat once more that a classical world, i.e. a world with $\hbar = 0$, would be devoid of any structure we know of.

We have already mentioned that QM synthesizes the contradictory concepts of wave and particle but at a cost which is twofold: (a) It introduces an abstract concept, the so-called wavefunction ψ, which is not easily reconciled with our perception of the World. (b) The required mathematical formalism goes clearly beyond the first two years of University calculus. What justifies this cost is the extraordinary explanatory power of QM which extends from the structure of protons and neutrons, etc. to the Whole Universe. Fortunately, one can retain a large fraction of this explanatory power without most of the cost. This can be achieved, as we shall show in this book, by: (a) Focusing on three basic principles of QM, namely those of Heisenberg, Pauli, and Schrödinger, which can be deduced directly from the wave-particle duality and (b) By employing the method of dimensional analysis.

3.2 Heisenberg's Uncertainty Principle[3]

The product of the uncertainty[4] Δx in the position of a particle (along a given direction) times the uncertainty Δp_x of the momentum component along the same direction is never smaller than $\hbar/2$:

$$\boxed{\Delta x \cdot \Delta p_x \geq \hbar/2}. \tag{3.3}$$

The physical meaning of (3.3) is the following: If a waveparticle is confined in a straight segment of length Δx, the component of its momentum p_x in the same direction cannot be exactly zero[5]; actually the average value of p_x^2 is larger than $\hbar^2/4\Delta x^2$. In other words, a waveparticle confined within a length Δx cannot be at

[3] Heisenberg's principle stems from the well known wave relation stating that the confinement of a wavepacket within a spatial extent Δx requires the superposition of a range of k's of extent $\Delta k_x \approx 1/\Delta x$; combining this relation with the wave-particle equation (3.2) we obtain essentially (3.3).

[4] Δx is the so-called standard deviation defined by the relation: $\Delta x^2 \equiv \left\langle (x - \langle x \rangle)^2 \right\rangle = \langle x^2 \rangle - \langle x \rangle^2$. A similar definition applies to Δp_x with x replaced by p_x The symbol $\langle f \rangle$, for any quantity f, denotes its average value.

[5] Although the average value of p_x can be zero, if the contributions of positive and negative values cancel each other.

rest; on the contrary, it must go back and forth perpetually as to satisfy the inequality $\langle p_x^2 \rangle \geq \hbar^2/4\Delta x^2$. In the realistic three dimensional (3D) case, where a particle may be confined within a sphere of radius r_o and volume V, we obtain[6] from (3.3):

$$\langle p^2 \rangle = \langle p_x^2 + p_y^2 + p_z^2 \rangle = 3\langle p_x^2 \rangle \geq \frac{3}{4}\frac{\hbar^2}{\Delta x^2}, \tag{3.4}$$

where

$$3\Delta x^2 = \langle x^2 + y^2 + z^2 \rangle = \langle r^2 \rangle = \frac{3}{5} r_o^2 \approx 0.23\, V^{2/3}. \tag{3.5}$$

Combining (3.4) and (3.5) with the expression for the non-relativistic[7] kinetic energy, we obtain for the average value of the latter

$$\boxed{\varepsilon_k = \langle p^2 \rangle / 2m \geq 4.87 \frac{\hbar^2}{mV^{2/3}} = 1.875 \frac{\hbar^2}{m\, r_o^2}} \tag{3.6}$$

Equation 3.6 is very important, because it accounts for the non-zero kinetic energy necessary for the stabilization of the various structures of the World. It asserts that a waveparticle confined within a volume V has a kinetic energy which cannot be smaller than a minimum non-zero value. This value is inversely proportional to the 2/3 power of the volume V, inversely proportional to the mass of the waveparticle, and proportional to the square of Planck's constant $\hbar$, the "trade mark" of QM. Thus, the smaller the volume and the lighter the mass, the larger the minimum value of the average kinetic energy is. If $\hbar$ were zero, as in classical physics, this minimum value would be zero and the World would be left unprotected against the collapse, which the attractive forces, left alone, would bring on. Notice that the minimum kinetic energy becomes zero, if $V \to \infty$, i.e., if the waveparticle is spatially unconfined in all directions.

It is important, as we shall see in future chapters, to calculate how (3.6) is modified in the extreme relativistic limit, $\varepsilon_k = cp$. Taking into account that $p \sim \hbar/\Delta x \sim \hbar/V^{1/3}$, we obtain:

$$\boxed{\varepsilon_k \geq a \frac{c\,\hbar}{V^{1/3}} \sim \frac{c\,\hbar}{r_o}, \quad a \approx 3} \tag{3.7}$$

Notice in (3.7) that the minimum kinetic energy is now inversely proportional to the one-third power of the volume within which the waveparticle is confined, or inversely proportional to its radius r_o. This change in the exponent of V is crucial in explaining the transition from white dwarfs to neutron stars and the collapse of neutron stars to black holes.

---

[6] By using the symmetry of the sphere and by assuming uniform density for any value of r.

[7] The correct relativistic relation between kinetic energy and momentum is $\varepsilon_k = \left(m_o^2 c^4 + c^2 p^2\right)^{1/2} - m_o c^2$. In the non-relativistic limit, $m_o c^2 \gg c\,p$, becomes $\varepsilon_k \approx p^2/2m_o$, while in the extreme relativistic limit, $m_o c^2 \ll c\,p$, becomes $\varepsilon_k = cp$, where m_o is the rest mass of the particle.

3.3 Pauli's Exclusion Principle

Identical waveparticles sharing the *same* volume V as they go on with their perpetual motion are *indistinguishable*. This feature is due to their wave-like motion and, consequently, the absence of trajectory; the latter, if it were present, they would allow us to tell which particle is which. The indistinguishability means that all physical properties must remain unchanged, if we interchange the "names" of any two of the identical particles. The absolute value of the square $|\Psi|^2$ of the wavefunction is such a physical property; hence, $|\Psi(1,2,\ldots)|^2$ must be equal to $|\Psi(2,1,\ldots)|^2$, where $\Psi(1,2,\ldots)$determines the state of a system consisting of several waveparticles and 1, 2, ... stand for the position and the direction of the spin of waveparticle no. 1, no. 2, etc. It follows that

$$\Psi(1,2,\ldots) = \pm\Psi(2,1,\ldots). \tag{3.8}$$

It turns out that the upper sign is valid for *all* waveparticles (elementary or not) having integer spins and the lower one for *all* waveparticles (elementary or not) with spin integer plus 1/2. The particles in the first category are called *bosons* and the ones in the second *fermions*. Thus all ic-waveparticles (such as photons, gluons, etc., see Table 2.2) are bosons and all elementary *m*-waveparticles (such as electrons, neutrinos, quarks, see Table 2.1) are fermions; protons and neutrons, consisting of three quarks are necessarily fermions. Consider now two identical fermions sharing the same volume and being in the *same* spatial/spin single-particle state: $\Psi(1,2) = \psi(1)\psi(2)$; if we interchange their names their joint wavefunction will remain the same, since the two particles are in the same single-particle state; however, (3.8) tells us that the joint wavefunction must change sign. The only way to satisfy both requirements is to set the joint wavefunction equal to zero, which means physically that the probability of finding two identical fermions in the same spatial/spin single-particle state is zero. Thus, the famous Pauli's exclusion principle was deduced:

> Out of many identical fermions, no more than one can occupy the same spatial/spin single-particle state. For 1/2 spin identical fermions (e.g., electrons, or protons, or neutrons) no more than two can occupy the same single-particle spatial state, one with spin up and the other with spin down. Less accurately, we can state that no more than two identical spin 1/2 fermions can be in the same region of space.

What will happen when N spin 1/2 identical fermions are forced to share the same homogeneous space of volume V? We can obtain a satisfactory answer to this question, if we assume that the identical spin 1/2 fermions divide the available space into equal subspaces each of volume $V/(N/2) = 2\ V/N$, and each pair of them (one partner of the pair with spin up and the other with spin down) are placed in each subspace; this way Pauli's principle is satisfied. Then, the average kinetic energy per fermion, assuming its non-relativistic expression, is obtained by replacing in (3.6) V by $2\ V/N$. We have then for the average kinetic energy per particle $\varepsilon_k \geq 4.87\,\hbar^2/m(2V/N)^{2/3} = -3.07\hbar^2 N^{2/3}/mV^{2/3}$. The average total kinetic energy of the N fermions, E_k, is larger than or equal to $N\varepsilon_k$:

$$\boxed{E_k \geq N \times 3.07\hbar^2 N^{2/3}/mV^{2/3} = 3.07\hbar^2 N^{5/3}/mV^{2/3};\quad 3.07 \rightarrow 2.87} \qquad (3.9)$$

An accurate calculation of the minimum value of the total kinetic energy produces the same result as in (3.9) with the numerical factor 3.07 replaced by 2.87.

We must draw attention to (3.9) which (under the said conditions of identical fermions sharing the same homogeneous space) boosts the minimum average kinetic energy per fermion (as compared with the Heisenberg-based (3.6)) by a factor proportional to $N^{2/3}$! This can be an enormous boost, if N is very large as in the case of "free" electrons in metals, where $N^{2/3}$ is of the order of 10^{15}! This enormous boost is absolutely necessary in order to stabilize e.g., condensed matter at the observed values of density, compressibility, etc., as we shall see in a future chapter.

3.4 Schrödinger's Principle of Spectral Discreteness

It is well known that that a classical wave confined within a finite region of space acquires a discrete frequency spectrum, as opposed to the continuous spectrum it possesses, if its extent is infinite (see Fig. 3.1b and a, p. 23). A very familiar example is a guitar string, which can produce only the fundamental frequency and its harmonics i.e., only the frequencies $\omega_n = (\pi\upsilon/\ell)n$, resulting from the general relation $\omega = (2\pi/\lambda)\upsilon$ and the requirement of an integer number of half wavelengths to fit exactly within the length ℓ of the string, $n(\lambda/2) = \ell$; n is a positive integer and υ is the velocity of the wave propagating along the string. Another example is the discrete spectrum of E/M waves confined within a cavity.

Because of the relation $\varepsilon = \hbar\omega$, we expect that a waveparticle, confined within a finite region of space, will acquire a discrete energy spectrum, in contrast to the continuous energy spectrum would have according to the classical worldview. In this book we will associate this wave-due discreteness with the name of Schrödinger, the discoverer of the equation which correctly describes the motion of non-relativistic waveparticles. We can restate Schrödinger's discreteness principle as follows: A waveparticle, confined within a finite region of space, can only have a series of allowed energies ε_o, ε_1, ε_2, ... These can be arranged in order of increasing magnitude with no other energy in between two successive allowed values (see Fig. 3.1(d′)). In information language, the energy of a confined particle is "digital" not "analog". We already know that ε_o is higher than the chosen zero level of energy by an amount equal to or larger than the minimum kinetic energy given by (3.6). It is also true that the difference $\delta\varepsilon \equiv \varepsilon_1 - \varepsilon_o$, in the non-relativistic case, is of the same order of magnitude:

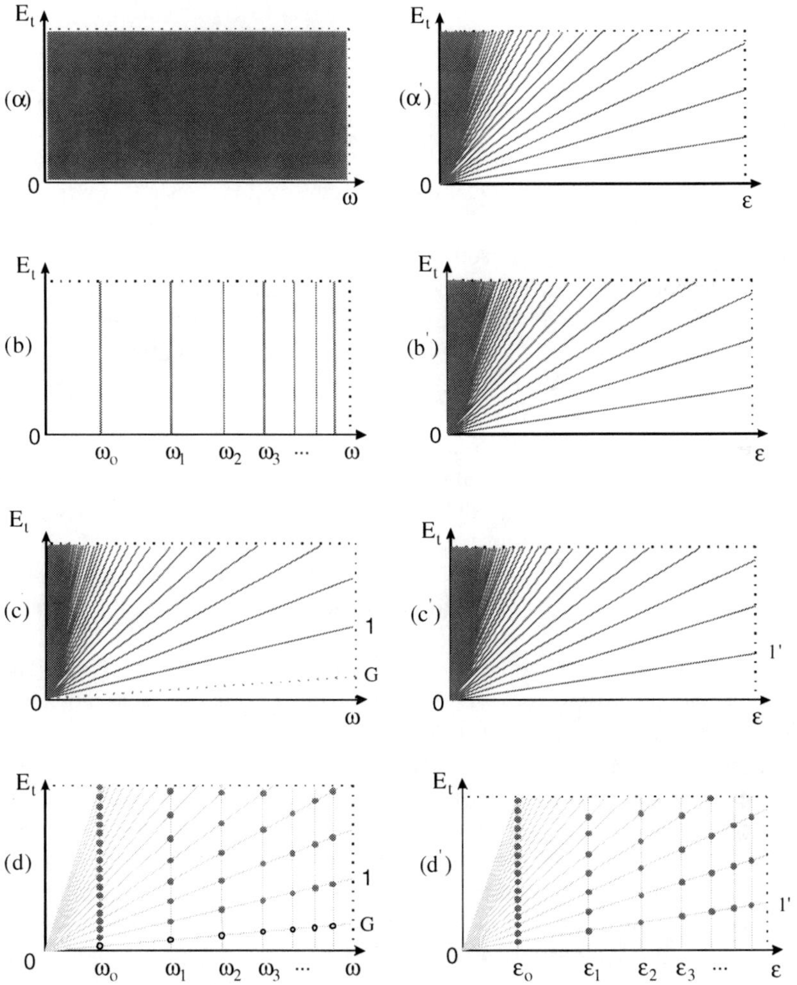

Fig. 3.1 **a–d** Allowed values in the plane, total energy E_t versus frequency ω: **a** unconfined classical wave; all points in the plane E_t vs. ω are allowed; **b** confined classical wave; only the points on the discrete lines perpendicular to the frequency axis are allowed (frequency discretization); **c** unconfined particlewave; only the points on the lines $E_t = n\hbar\omega$ are allowed where $n = 1, 2, 3,\ldots$ (total energy quantization); **d** Confined particlewaves; only the points marked by dots are allowed (frequency discretization and total energy quantization). **a′–b′** Allowed values in the plane, total energy, E_t versus particle energy ε. **a′** Unconfined n ($n = 1, 2, 3,\ldots$) non-interacting classical particles, each of the same ε. **b′** As in (**a′**) but confined. **c′** Unconfined n ($n = 1, 2, 3,\ldots$) non-interacting waveparticles, each of energy ε. In cases (**a′**), (**b′**), (**c′**) only the points on the lines $E_t = n\varepsilon$ are allowed. (**d′**) As in (**c′**), but confined; discretization along the E_t axis (as in cases (**a′**), (**b′**)), due to particle aspects of waveparticles and quantization along the ε-axis (as a result of the wave aspect of waveparticles). Strictly speaking, only cases (**c**), (**d**), (**c′**), (**d′**) correspond to the reality, which is of quantum nature. Notice that, as ω or ε tend to zero, the classical and the quantum behavior tend to coincide

$$\delta\varepsilon = c_1 \frac{\hbar^2}{m\,V^{2/3}} \tag{3.10}$$

where the numerical factor c_1 is of the order of one; its exact value depends on the type of the potential responsible for the confinement. Equation 3.10 is of similar crucial importance as the combination of Heisenberg's and Pauli's principles leading to (3.9). Equation 3.9 guaranties the stability and, hence, the existence of the World. Equation 3.10 implies than any temporary perturbation of size smaller than $\delta\varepsilon$ will leave a waveparticle—even a composite one—unchanged in its ground state corresponding to the energy ε_o, if it were in this state initially. This means that composite waveparticles, or even systems, behave as elementary ones up to a limit. In other words, composite waveparticles, although they possess the possibility of change—even breaking—they require for this to happen the application of an external perturbation *exceeding* a non-zero value, which, in some cases, can be much larger than any available perturbation. It is exactly this property which allows us to attribute a predictable chemical and physical behavior to e.g., any atom of the periodic table, in spite of being under the influence of quite different environments. In contrast, in a classical world, where $\hbar$ would be zero, the behavior of an atom would never be the same, because the ever present interactions with its environment-even the most minute ones-would force the atoms to be in a state of continuous change. Equation 3.10 shows that the smaller a composite waveparticle is, the more stable it is. This explains why non-radioactive nuclei cannot change under ordinary conditions and why the alchemist had no chance to change other common substances to gold.

The picture of the ground state energy and the next excited energy level is different for waveparticles, such as electrons, than for particlewaves, such as the quanta of the EM field. This difference stems from the fact that the number of electrons is conserved, while the number of photons is not. Thus, we can talk about the number of electrons in a system and its energy as independent variables. E.g., we can consider just one electron being in its ground state (of energy ε_o in case of Fig. 3.1(d′), or 0 in case of Fig. 3.1(c′)). *Any excitation of this one-electron system moves along the straight line* 01′ *of Fig. 3.1*(c′)*, or by jumping to the next point in the sequence of points lying on the straight line* 01′ *of Fig. 3.1*(d′). In contrast to this picture, the number of photons is not an independent variable; its value is determined from the value of the total energy (and ω, if the latter is given). In particular, the ground state energy, E_G (i.e., the lowest possible energy) of the EM field involves zero number of photons, but it is not zero; it consists of the sum of all points on the line 0G (see Fig. 3.1c or d). This is due to the fact that, according to quantum mechanics, the electric field $\boldsymbol{E}$ and the magnetic field $\boldsymbol{B}$ cannot be simultaneously exactly zero, for the same reason that δx and δp_x cannot be zero simultaneously. In other words the sum of $\boldsymbol{E}^2 + \boldsymbol{B}^2$, which is proportional to the energy density of the electromagnetic field, cannot be exactly zero; its lowest value is equal to the sum of all $\frac{1}{2}N_i\hbar\omega_i$, where N_i is the number per unit volume of all the waves of frequency ω_i; for dimensional reasons $N_i \propto \omega_i^3/c^3$ (see Chap. 5). In other words, the quantum "vacuum", as far as the EM field is concerned, is not a completely empty space, but contains the unavoidable minimum fluctuations of the EM field vectors around their average value; the latter is zero, because of cancellation of positive and negative values. *The lowest excited energy level of the EM field for a given allowed ω is to move vertically from a point in the line 0G to the point on the line 01 (see Fig. 3.1c or d).* The energy of this state is $E_G + \hbar\omega$ and it involves just one photon of frequency ω. As it is clear from Fig. 3.1c or d, the higher the frequency, the higher the minimum excitation energy $\delta\varepsilon = \hbar\omega$ is. This means that we need higher external energy to excite a photon of higher frequency. The most common way to excite a system from its ground state is to bring it in contact with a heat bath of absolute (Kelvin) temperature T; then, the available energy that the heat bath can possibly transfer to the system is within a range from 0 up to a maximum value which is of the order k_BT, where k_B is the so-called Boltzmann's constant.[8]

Thus, a heat bath of temperature T in equilibrium with a photon system cannot appreciably excite photons of frequency higher than $\omega_o = c_1 k_B T/\hbar$, where c_1 is a constant of the order of one. This QM limit of the highest excitable frequency of a photonic system finds immediate application in calculating the total EM energy, I, emitted by a black body per unit time and per unit area. The quantity I can be obtained by integrating over ω the quantity $I_\omega d\omega$ corresponding to the frequency range [ω, $\omega + d\omega$]. The classical result for $I_\omega d\omega$ is proportional to $(k_BT/c^2)\omega^2 d\omega$ as we shall see in Chap. 5. If this quantity is integrated from zero to infinity, as in classical physics, one gets infinity, an absurd result. If the upper limit of integration is restricted to the QM one, $\omega_o = c_1 k_B T/\hbar$, the result will be proportional to $\frac{1}{3}(k_BT/c^2)\,\omega_o^3 = \left(c_1^3/3\right)(k_BT)^4/c^2\,\hbar^3$ in agreement with the experimental data.

We shall conclude this chapter by mentioning an alternative picture, proposed by Feynman, of how a waveparticle propagates from a point A to a point B. If it were a classical particle, it would follow a single unique trajectory from A to B. In contrast, the waveparticle will follow *all possible trajectories* joining the two points, each trajectory weighted with what is called a probability amplitude. The absolute value square of the sum of all these probability amplitudes gives the probability for the waveparticle to propagate from A to B. Usually, to a good approximation, the trajectories which give an appreciable contribution to the sum are those which are within a tube of cross-section proportional to the wavelength square around the classical trajectory. Feynman's alternative formulation of quantum mechanics, called path integral formulation [3.8], is equivalent to the formulation through ψ obeying Schrödinger's equation. Conceptually, it is probably easier to reconcile our perceptions of particle motion with Feynman's approach rather than with the ψ formulation, since the former still makes use of the trajectory concept, although not a single one. Computationally, it is usually simpler to work with ψ, especially for bound systems, as compared with the path integral formulation, although there are cases where the latter is more convenient.

---

[8] For the numerical values of all universal constants see Table 1 in p. 141.

Chapter 4
Equilibrium and Minimization of Total Energy

Abstract The First and the Second Laws of Thermodynamics are presented. Their combination leads to the conclusion that one of the thermodynamic potentials, such as the energy or the free energy, becomes minimum when thermodynamic equilibrium is reached. Which one of them is minimized at equilibrium is determined by the boundary conditions. The important concept of chemical potential is introduced.

Keywords Equilibrium · Minimization of energy · Entropy · Availability · Gibbs free energy · Second law

Summary The stable equilibrium state of a physical system of negligible absolute temperature, under conditions of receiving no work and no m-particles from the environment, corresponds to the minimum of its total energy U.[1]

Under the more common conditions of a physical system being in contact with an environment of constant temperature T_o and constant pressure P_o, the system's stable equilibrium state corresponds to the minimum of a quantity A called *availability* and defined as follows:

$$A \equiv U + P_o V - T_o S \tag{4.1}$$

U is the total energy[1] of the system, V is its volume, and S is its entropy. These minimization principles stem from the 2nd law of thermodynamics, i.e., the law of

---

[1] More accurately, U is the so-called internal energy, which is defined as the average value of the total energy E_t of the system, under conditions of the total momentum being equal to zero on the average. (For macroscopic systems, the total angular momentum must be zero on the average as well). Notice that the value of U or E_t is fully determined only after we choose a reference state, the energy of which is by definition zero. Usually, the reference state is the one at which all the particles of the system are at infinite distance from each other and at their ground state. There are three types of contributions to U or to E_t: The rest relativistic energy $E_o = \sum m_{oi}c^2$, the kinetic energy E_K, and the potential energy E_P, which may include interactions both with the environment and among the particles of the system itself. For convenience, it is not uncommon to ignore the rest energy by incorporating it in the reference state, if no changes in it are involved.

E. N. Economou, *A Short Journey from Quarks to the Universe*, SpringerBriefs in Physics, 1, DOI: 10.1007/978-3-642-20089-2_4,

entropy increase. Expanding $U(V, S)$ in powers of V and S up to second order and replacing in (4.1), we find that the minimization of A implies that $P = P_o$ and $T = T_o$ as well as some general thermodynamic inequalities.[2] Thus, at equilibrium, the availability A coincides with *Gibbs free energy*, $G \equiv U + PV - TS$

In this short chapter the 1st and the 2nd law of thermodynamics are presented together with the concepts of entropy and the Gibbs free energy. Their central role for equilibrium structures is pointed out (For more details see Ref. [4.1]).

4.1 The First Law

The first law is essentially the law of conservation of energy. It states that the infinitesimal change dU of the energy U of a system is due to the following three exchanges with the environment: (a) receiving an infinitesimal quantity of heat $đQ$; (b) giving an infinitesimal work $đW$; and (c) receiving an infinitesimal amount of energy $đE_m$ associated with the inflow of an infinitesimally small number of m-particles. Then, conservation of energy implies that

$$\boxed{dU = đQ - đW + đE_m} \tag{4.2}$$

The infinitesimal quantities $đQ$, $đW$, $đE_m$ are not perfect differentials of certain functions Q, W, E_m, respectively; such functions do not exist. We cannot separate the internal energy of a system into these three components. Only during the process of *exchange* of energy such a separation makes sense for the energy being exchanged. Mathematically speaking, $đQ$, $đW$, $đE_m$ are the so-called differential forms and not perfect differentials. It is exactly this distinction which is denoted by the symbol $đ$ instead of d in front of Q, W, E_m.

The second law, one of the most basic laws of Nature connecting the macroscopic world to its microscopic constituents, can be stated in several equivalent ways. One of the most revealing is through the concept of entropy. The latter is only defined in systems consisting of so many microscopic particles, that an average macroscopic description of these systems becomes imperative. The entropy of such a system is proportional to the logarithm of the total number, $\Gamma(U, V, N, \ldots)$, of microscopic states of the system corresponding to the macroscopic state fully described by the independent variables[3] $U, V, N, \ldots$:

$$\boxed{S \equiv k_B \ln \Gamma(U, V, N, \ldots),} \tag{4.3}$$

---

[2] These inequalities are: $T \geq 0$, $C_V > 0$, $C_P > C_V$, $(\partial P/\partial V)_T < 0$.

[3] A simple system, such as a perfect gas, when *in equilibrium*, can be described macroscopically by only three independent macroscopic variables U, V, N; for a photon system in equilibrium N is not an independent variable and, hence, only two independent variables are sufficient for its macroscopic description (see p. 35). Other more complicated systems in equilibrium may require a larger number of independent macroscopic variables. Non-equilibrium states of a system require more independent macroscopic variables than the ones required for the equilibrium state of the same system.

4.2 The Second Law

The second law states that the entropy of a system, *which does not exchange heat and m-particles with the environment,* is always increasing until the system reaches the equilibrium state at which the entropy has its maximum value. Thus, for *such an isolated* system the maximum entropy implies equilibrium and vice versa. From the 2nd law, it follows that processes taking place within the system in its way towards equilibrium increase its entropy. Hence, the reverse processes, if they could happen without external intervention, they would decrease the entropy, they would violate the 2nd law, and as such they are impossible. We conclude that processes driving an isolated system towards equilibrium are *irreversible*; only in extreme limiting cases such a process can be considered as reversible, and only if it leaves the entropy unchanged. Examples of irreversible processes are: Heat transfer among parts of the system at different temperatures, diffusion of particles among parts of the system at different concentrations, chemical reactions within the system, etc.

What does it happen when a system exchanges heat with the environment but no m-particles? In this case the entropy changes in two distinct ways: The internal one, as before through irreversible processes taking place within the system, and the external one due to the exchange of heat.

$$\boxed{dS = đS_i + đS_e, \quad \text{where } đS_i \geq 0,\ đS_e = đQ/T \text{ and } đE_m = 0} \tag{4.4}$$

From (4.4) and the positiveness of temperature at equilibrium, another equivalent version of the second law follows:

$$TdS \geq đQ \tag{4.5}$$

Combining (4.2) and (4.5), i.e., the 1st and the 2nd law, we obtain:

$$\boxed{dU \leq TdS - đW + đE_m} \tag{4.6}$$

Equation (4.6) shows that, when $Tds = đW = đE_m = 0$, $dU \leq 0$; this means that, under the conditions $Tds = đW = đE_m = 0$, U keeps decreasing until the equilibrium state is reached, at which the energy becomes minimum. Assuming that $đW = PdV$ and recalling the definition of the Gibbs free energy, $G = U + PV - TS$, we obtain, taking into account (4.6):

$$\boxed{dG \leq -SdT + VdP + đE_m} \tag{4.7}$$

From (4.7), under conditions of constant temperature and pressure and no exchange of m-particles, it follows that $dG \leq 0$; this means that the Gibbs free energy, under the said conditions, is always decreasing in its way towards equilibrium (only in extreme limiting cases may remain unchanged) and it reaches its minimum value at equilibrium.

4.3 Maximum Work, Gibbs' Free Energy, and Chemical Potential

Let us consider now a system *not in equilibrium* and in contact with an environment of constant temperature T_o and constant pressure P_o. What is the maximum work, W_M, which can be extracted from such a system by exploiting its non-equilibrium state? It turns out [4.2] that this maximum work is obtained only in the limiting case where the total entropy of the system and the environment together remains unchanged during the whole process; then W_M, is given by the formula

$$W_M = A_{\text{initial}} - A_{\text{final}} \tag{4.8}$$

where A is the availability defined in (4.1). If $P = P_O$ and $T = T_O$ then the maximum work is

$$W_M = G_{\text{initial}} - G_{\text{final}} \tag{4.9}$$

Equations (4.8) and (4.9) provide a justification for the names availability and free energy for A and G, respectively.

As an elementary application of (4.6), let us minimize $U = E_K + E_P$ with respect to the volume V of the system, under the conditions $T_o = P_o = đE_m = 0$. We have

$$(\partial U/\partial V) = (\partial E_K/\partial V) + (\partial E_P/\partial V) = 0 \tag{4.10}$$

However, at equilibrium, $(\partial U/\partial V) = -P$, $(\partial E_K/\partial V) = -P_K$ and $(\partial E_P/\partial V) = -P_P$. Thus, the minimization of the energy U with respect to the volume, under the conditions, $T_o = P_o = đE_m = 0$ implies the obvious relation that, at equilibrium, $P_K = -P_P$, which means that the expanding pressure of the kinetic energy equals to the absolute value of the squeezing pressure of the potential energy.

For processes starting from an initial equilibrium state and ending at another equilibrium state, while keeping $đS_i = 0$ during the whole duration of the process, the equality sign in (4.6) and (4.7) holds. For such processes we have:

$$(\partial U/\partial S)_{V,N} = T, \quad (\partial U/\partial V)_{S,N} = -P, \quad (\partial U/\partial N)_{S,V} = \mu, \tag{4.11}$$

$$(\partial G/\partial T)_{P,N} = -S \quad (\partial G/\partial P)_{T,N} = V, \quad (\partial G/\partial N)_{T,P} = \mu \tag{4.12}$$

where the subscripts indicate which variables are kept constant during the differentiation. The quantity μ, called the *chemical potential*, is a very important one, because at equilibrium it must have the same value throughout the system, similarly to the temperature. Moreover, in non-equilibrium states where μ may have different values at different parts of the system, particles flow from regions of high μ to regions of low μ until equalization of μ is achieved. Finally, it is easy to show that μ is simply related to Gibbs' free energy: $G = N\,\mu(T, P)$. For photons, as well as for any other type of particles whose number is not conserved, the chemical potential is zero; this follows from the fact that the number of such particles is not an independent variable and, hence, the term $\text{đ}E_m = \mu\, dN$ must not be present in (4.6) and (4.7).

The internal energy U appearing in (4.11) can be calculated from (4.3) by solving U as function of S, V, N, the "natural" independent variables for U. The corresponding "natural" independent variables for G are the following: T, P, N. In order to change independent variable from S to T, and from V to P we added to U the products $-TS$ and PV, respectively, so that $G \equiv U - TS + PV$ resulted. In other words, the definition of G is coming out of the requirement of having T, P, N as independent "natural" variables. If we wanted a quantity having as independent "natural" variables the set T, V, N we must add to U the product $-TS$ to define the so-called *Helmholtz free energy*, $F \equiv U - TS$. For computational purposes, it is needed sometimes to define a thermodynamic potential, denoted by the symbol Ω with T, V, μ as "natural" independent variables. We have then that $\Omega \equiv F - \mu N$. It is easy to show that $\Omega = -PV$.

We conclude by emphasizing the main point of this short chapter: We possess a recipe to predict (or to justify a posteriori) the equilibrium structures of matter and their properties. Here is the recipe:

For the system under study obtain the Gibbs free energy G (under conditions of constant P and T and $\text{đ}E_m = 0$), or the energy U (under conditions $P = T = 0$ and $\text{đ}E_m = 0$), as a function of various other free parameters (such as the volume, the positions of the atoms or ions, the electronic concentration $n_e(\mathbf{r})$, etc.); minimize G or U with respect to all these free parameters; the resulting state is one of stable equilibrium to be observed in nature.

Chapter 5
Dimensional Analysis: A Short-cut to Physics Relations

Abstract The method of dimensional analysis is presented and is applied to derive properties of photons and of photonic interactions with matter. It is also employed in the derivation of certain relations concerning oscillatory and wavy motions.

Keywords Dimensional analysis · Photons · Eigenfrequencies · Wave velocity

Outline of the Method Under certain circumstances, the dimensional analysis allows us to find out, how a given quantity X depends on the "relevant" independent variables and parameters. As the reader may have guessed, the crucial and far from trivial task is to identify all the "relevant" quantities on which the value of the given quantity may depend. Once this task has been achieved, the next step is to find the more general combination of the "relevant" quantities having the same dimensions (i.e., the same powers of length, time, and mass) as the given quantity. To be more specific the dimensional analysis works as follows:

1. We identify all possible quantities $A_1, \ldots A_n$ on which the quantity of interest, X, may depend.
2. If $n \leq 3$, there is usually a unique combination of the form $A_1^{\mu_1} \ldots A_n^{\mu_n}$ which has the same dimensions as X. Then the desired formula for X has been found: $X = \eta A_1^{\mu_1} \ldots A_n^{\mu_n}$ where η is an undetermined numerical constant.
3. If $n > 3$ we choose three quantities A_1, A_2, A_3 which define a system of units (in the sense that three different combinations of the form $A_1^{\mu_1} A_2^{\mu_2} A_3^{\mu_3}$ can be found which have the dimensions of length, time, and mass, respectively).
4. By a proper choice of ν_1, ν_2, ν_3 we form the combination $A_1^{\nu_1} A_2^{\nu_2} A_3^{\nu_3} \equiv X_o$, where X_o has the same dimensions as the quantity of interest X.
5. We create also combinations $A_1^{\xi_{1n}} A_2^{\xi_{2n}} A_3^{\xi_{3n}} \equiv A_{no}$ which have the same dimensions as A_n, ($n = 4, 5, \ldots$) and we define the dimensionless quantities A_n/A_{no} ($n = 4, 5, \ldots$).

E. N. Economou, *A Short Journey from Quarks to the Universe*,
SpringerBriefs in Physics, 1, DOI: 10.1007/978-3-642-20089-2_5,

6. Having determined the quantities X_o, A_{40}, A_{50},... in terms of the chosen quantities A_1, A_2, A_3, we are ready to express X in terms of $A_1, \ldots A_n$ as follows:

$$X = X_0 f\left(\frac{A_4}{A_{40}}, \frac{A_5}{A_{50}}, \ldots\right)$$

where the unknown function f cannot be determined from dimensional analysis; additional information(s) or even a complete physical theory is needed to find out what f is.

Several examples of the method of dimensional analysis are presented in this chapter demonstrating its extraordinary power.

5.1 Photons in Equilibrium

How the emitted E/M energy, I_ω per unit time, per unit frequency, and per unit area from a black body is distributed among the various frequencies?

A black body is one which absorbs 100% of the E/M radiation falling on it; the emission properties of such a body are characterized by only one parameter, its absolute temperature. However, keep in mind that in Nature the absolute temperature appears always as the product $k_B T$ with dimensions of energy, where k_B is a universal constant called Boltzmann constant. Thus, one quantity on which I_ω depends is the combination $k_B T$; another is, obviously, the frequency ω; finally, the velocity of light c is expected to appear in I_ω, since it is the main quantity characterizing the E/M radiation. From the point of view of classical physics these three quantities are the only ones determining I_ω. The dimensions of I_ω are energy over length square. To obtain the same dimensions from the combination $(k_B T)^{\mu_1} \omega^{\mu_2} c^{\mu_3}$ we must choose $\mu_1 = 1$, $\mu_2 = 2$, $\mu_3 = -2$. Hence, the classical result for I_ω is

$$I_\omega = a\,(k_B T)\omega^2/c^2 \tag{5.1}$$

where a is a numerical factor which turns out to be equal to $1/(4\,\pi^2)$. The result (5.1) coincides with the formula we have used at the end of Chap. 3 (p. 25). It is clear that the classical result (5.1) cannot be right for high frequencies ($\hbar\omega \geq k_B T$), since its integral over all frequencies would produce an infinite result for the total radiation emitted by a black body. There are two solutions around this problem: One is the introduction of the cut-off we employed in Chap. 3; the other is to extend the integration over all frequencies but with a I_ω incorporating the correct quantum behavior at all frequencies (see Fig. 3.1, p. 23), which means that a fourth quantity, that of $\hbar$, will enter in the expression for I_ω. Thus the correct formula for I_ω is the following:

$$I_\omega = a\,(k_B T)\left(\omega^2/c^2\right)f(\hbar\,\omega/k_B T) \tag{5.2}$$

where f is a function which cannot be obtained by dimensional analysis. We know though that (5.2) must reduce to (5.1), when $\hbar\,\omega/k_B T \ll 1$, so that $f \to 1$, for

$\hbar\omega/k_BT \to 0$; moreover, in the opposite limit, $\hbar\omega/k_BT \gg 1$, f must approach zero. It turns out that $f(\hbar\omega/k_BT) = (\hbar\omega/k_BT)\left[e^{\hbar\omega/k_BT} - 1\right]^{-1}$, [4.2].

Employing dimensional analysis determine the total E/M energy I emitted by a black body of temperature T per unit time and per unit area.

The quantity I depends on the following three quantities: The combination k_BT, the velocity of light c (as an E/M phenomenon), and Planck's constant $\hbar$ (as an explicitly quantum phenomenon). The combination $(k_BT)^{\mu_1}\hbar^{\mu_2}c^{\mu_3}$ which has the same dimensions as I, i.e., energy over time and over length square is obtained only if $\mu_1 = 4$, $\mu_2 = -3$, and $\mu_3 = -2$. These values become obvious by taking into account that $\hbar/k_BT$ has dimension of time and $\hbar c/k_BT$ has dimension of length. Hence

$$I = a'(k_BT)/(\hbar/k_BT)(\hbar c/k_BT)^2 = a'(k_BT)^4/\hbar^3c^2 \tag{5.3}$$

where integration of (5.2) over all frequencies and taking the outward normal component of the flow shows that the numerical factor is $a' = \pi^2/60$.

Employing Dimensional Analysis Determine the Thermodynamic Quantities of a Photon Gas in Equilibrium.

As it was discussed in Chap. 4 (p. 27), only two independent thermodynamic variables determine the equilibrium properties of a photon gas: One is the volume V and the other is the temperature T. Moreover, due to the extensive character of the thermodynamic potentials (such as the energy, the free energy, etc.), all these quantities must be proportional to the volume. E.g., $U = V\varepsilon$, where by definition ε is the energy density with dimensions energy over length to the third power. The energy density ε must depend on the combination k_BT, on the velocity of light c (since it is an E/M quantity), and on $\hbar$ (since photons are the quanta of the E/M field). Taking into account that $c\hbar/k_BT$ has dimensions of length, we conclude that $\varepsilon = a\,(k_BT)/(c\hbar/k_BT)^3 = a\,(k_BT)^4/c^3\hbar^3$; a detailed calculation shows that the numerical constant a is equal to $\pi^2/15$. Thus, the result for the energy is the following:

$$U = \left(\pi^2/15\right) V (k_BT)^4/c^3\hbar^3 \tag{5.4}$$

Taking into account (5.4) and that $(\partial U/\partial S)_V = T$, we obtain for the entropy S that $S = \frac{4}{3}U/T$; then Helmholtz' free energy is $F = U - TS = -\frac{1}{3}U$ and the pressure P is $P = -(\partial F/\partial V)_T = \frac{1}{3}\varepsilon = \frac{1}{3}(U/V)$. The Gibbs free energy is $G \equiv F + PV = -\frac{1}{3}U + \frac{1}{3}U = 0$, in agreement with the fact that the chemical potential, $\mu \propto G$, is zero for the photon gas, due to the non-conservation of the number of photons.

5.2 Emission of Photons

Radiation by a Moving Particle of Electric Charge q.

The electromagnetic (E/M) energy per unit time J emitted by such a particle must depend on q (since the electric charge is the source and the receiver of E/M energy), on c (since the velocity of light is characterizing the E/M radiation), and on the acceleration, a, of the particle (since, from all kinematic quantities only a is directly connected to the force acting on the particle and providing the necessary work to compensate for the emitted energy). There is only one combination of these three quantities having dimensions of energy over time: q^2a^2/c^3. Hence

$$J = \eta q^2 a^2 / c^3 \tag{5.5}$$

The numerical factor turns out to be 2/3, if a^2 is the time average of the acceleration square assumed of sinusoidal time dependence.

Radiation by a neutral system possessing an oscillating dipole moment $p = \sum q_i r_i$.

From (5.5) and taking into account the definition of the dipole moment and that $a = -\omega^2 r$ (if the oscillation is sinusoidal) we expect that

$$J = \eta p^2 \omega^4 / c^3 \tag{5.6}$$

To the same formula we would arrive by taking into account that J in this case depends on $\boldsymbol{p}$, ω, c

Use (5.6) *to obtain the Life-Time of an Excited State.*

The product of J times the life-time τ of an excited state gives the energy emitted during the time τ; but this energy is just the energy $\hbar\omega$ of the emitted photon, assuming that the deexcitation occurs by the emission of a photon through a dipole transition. Hence

$$\tau = \hbar\omega / J = \hbar c^3 / \eta p^2 \omega^3 \tag{5.7}$$

Applying (5.7) to the transition $2p \rightarrow 1s$ of the hydrogen atom, where $\hbar\omega$ is equal to 10.2 eV (see Chap. 8, in particular (8.1)) and taking $p^2 \approx e^2 a_B^2$ and $\eta \approx 2/3$ we find $\tau \approx 1.77 \times 10^{-9}$s versus $\tau = 1.59 \times 10^{-9}$s obtained by a detailed quantum mechanical calculation.

5.3 Scattering of Photons

Any scattering event is characterized quantitatively by the so-called total scattering cross-section, σ_s which can be thought of as the area of the incoming beam intercepted by each scatterer. More precisely, σ_s is defined as the ratio Γ/F, where Γ is the number of scattered incoming particles per unit time and F is the flux of these incoming particles. The flux is in general equal to the concentration of the incoming particles times their common velocity. In particular for photons, we have that the number of scattered photons per unit time is the emitted energy per unit time J over the energy of each photon, $\Gamma = J/\hbar\omega$; similarly, the concentration of photons is equal to their energy density, $E_o^2/4\pi$, divided by the energy $\hbar\omega$ of each photon. Hence, $F = \text{concentration of photons} \times c = c E_o^2 / 4\pi\hbar\omega$. Thus

$$\sigma_s = \frac{J}{\hbar\omega} \frac{4\pi\hbar\omega}{cE_o^2} = \frac{4\pi J}{cE_o^2} \tag{5.8}$$

For a charged particle, (5.6) gives $J = \frac{2}{3}q^2a^2/c^3$, where the acceleration a is connected to the electric field: $a = q\mathbf{E}_0/m$. If the scatterer is an electron, we have $q = e$ and $m = m_e$ thus

$$\sigma_s = \frac{8\pi}{3}\left(\frac{e^2}{m_e c^2}\right)^2 = 6.65 \times 10^{-29}\ \mathrm{m}^2 \tag{5.9}$$

where $e^2/m_e c^2 = r_e = 2.82 \times 10^{-15}$m is the classical "radius" of the electron (resulting by equating the rest energy $m_e c^2$ to the electrostatic energy e^2/r_e). Equation (5.9) can be obtained by dimensional analysis taking into account that the scattering cross-section σ_s must depend on q (for $q = 0$ there is no scattering), on m (for $m = \infty$ there is no motion of the particle and, therefore, no scattered radiation), on c, and on $\hbar\omega$. Thus, dimensional analysis predicts for σ_s the following expression (known as the Klein–Nishina formula)

$$\sigma_s = c_1\left(\frac{q^2}{m\,c^2}\right)^2 f\left(\frac{\hbar\omega}{m\,c^2}\right) \tag{5.10}$$

where the function f of the dimensionless ratio $\hbar\omega/m\,c^2$ cannot be determined by dimensional analysis. Actually, if c_1 is taken equal to $8\pi/3$, f is a decreasing function of $\hbar\omega/mc^2$ which is one for $\hbar\omega = 0$ and zero for $\hbar\omega/mc^2 \to \infty$.

The scattering cross-section of a photon by a neutral atom or molecule of zero average dipole moment is due to the *induced* dipole moment $\boldsymbol{p}$ by the external field E_o

$$p = a_p E_o \tag{5.11}$$

where a_p is the polarizability of the atom or the molecule. Substituting (5.6) and (5.11) in (5.8) we obtain

$$\sigma_s = \frac{8\pi\omega^4}{3\ c^4}a_p^2 \tag{5.12}$$

It follows from (5.11) that the polarizability a_p has dimensions of length cube

$$a_p \simeq c_1 r_o^3 f_R(\omega) \tag{5.13}$$

where c_1 is a numerical factor in the range roughly between 4 and 10, r_o is the radius of the atom (for the molecules N_2 and O_2, $a_p = 11.75a_B^3$ and $10.67a_B^3$ respectively) and $f_R(\omega)$ is a typical resonance function

$$f_R(\omega) \approx \frac{\omega_o^2}{\omega_o^2 - \omega^2 - i\omega/\tau} \tag{5.14}$$

which takes into account the strongly increased response of the atom (or molecule) when the frequency ω of the incoming EM field coincides with one of its eigenfrequencies $\omega_o = |\varepsilon_\mu - \varepsilon_\nu|/\hbar$. For $\omega \ll \omega_o$, $f_R(\omega) \simeq 1$, while for $\omega = \omega_o$

$$|f_R(\omega)| = \omega_o \tau \tag{5.15}$$

where τ is the lifetime given by (5.7). Thus, for $\omega \ll \omega_o$ and for the hydrogen atom for which $c_1 \approx 4.5$ we have by introducing $\lambda = 2\pi c/\omega$ and $\lambda_o = 2\pi c/\omega_o$ that

$$\sigma_s \approx 2.6 \times 10^5 \frac{a_B^6}{\lambda^4}, \quad \lambda \gg \lambda_o \approx 1200\,\mathrm{A} \tag{5.16}$$

For the atmospheric air the coefficient in (5.16) will be 1.72×10^6.

At the resonance, using (5.15) in combination with (5.7) and setting $p^2 = c_2 \hbar \omega_o r_o^3$ (since $\hbar\omega_o \approx e^2/r_o$), we obtain $|f_R(\omega)| = \omega_o \tau = c^3/\eta\, c_2 r_o^3 \omega_o^3$. Substituting this expression in (5.13) and then in (5.12) we get

$$\sigma_s = \frac{8\pi}{3}\left(\frac{c_1}{2\pi\eta c_2}\right)^2 \lambda_o^2, \quad \lambda = \lambda_o \tag{5.17}$$

For the hydrogen atom, where $c_2 \approx 2.66$, $c_1 \approx 4.5$, we obtain $\sigma_s \approx 1.36\lambda_o^2 \approx 2 \times 10^{-14}\,\mathrm{m}^2$.

Equation 5.16 means that the off-resonance scattering cross-section of an *optical* photon, $\lambda \approx 600\,\mathrm{nm}$, by a neutral N_2 molecule is about equal to $2.88 \times 10^{-31}\,\mathrm{m}^2$, i.e., about eleven orders of magnitude less than the geometrical cross-section πr_o^2. On the other hand, at the first resonance of the hydrogen atom, $\hbar\omega_o = 10.2\,\mathrm{eV}$, the scattering cross-section is, according to (5.17), six orders of magnitude larger than the geometrical cross-section!

Total Scattering Cross-Section and Mean Free Path.

Important transport properties, such as electrical and thermal conductivities in solids, as well as propagation of E/M waves depend on the so-called mean free path ℓ. For an electromagnetic wave propagating in the x direction within a medium, the mean free path is defined by the relation $E^2(x) = E^2(0)\exp(-x/\ell)$; if the medium is gaseous, the mean free path is connected to the total scattering cross-section σ_s by each scatterer as follows:

$$1/\ell \approx n_s \sigma_s \tag{5.18}$$

where n_s is the concentration of scatterers. Equation (5.18) is valid for weak scattering, i.e., when the mean free path is much larger than the wavelength of the propagating wave. For Earth's atmosphere at sea level the concentration of molecules can be found from the law of perfect gases, $n_s = P/k_B T = 2.4 \times 10^{25}\,\mathrm{m}^{-3}$, for $P = 1$ atm and $T = 300$ K. Using our previous estimate for σ_s, we obtain for the photonic mean free path $\ell \approx 140\,\mathrm{km}$.This estimate is an idealized upper limit; actually the mean free path of visible light in the lower atmosphere can be much smaller, because to the $n_s\sigma_s$ of (5.18) other similar terms must be added due to the concentration of dust particles and water droplets and their corresponding scattering cross-section.

5.4 Relations Regarding Some Eigenfrequencies and Wave Velocities

Determine by dimensional analysis the natural frequency (called eigenfrequency) of oscillation of a pendulum (Fig. I(a), p. 126) and of an LC circuit (Fig. I(b)).

For the pendulum case the relevant quantities are: The length l, the acceleration of gravity g (since it is the gravity that provides the potential energy), and possibly the mass m. The only set of values of μ_1, μ_2, μ_3 which will give to the combination $l^{\mu_1} g^{\mu_2} m^{\mu_3}$ dimensions of one over time (as that of frequency) is $\mu_1 = -\frac{1}{2}$, $\mu_2 = \frac{1}{2}$, $\mu_3 = 0$. Hence, $\omega = \sqrt{g/l}$, where a detailed calculation shows that the unavoidable numerical factor turns out to be equal to one in this case.

For the LC circuit the obvious relevant quantities are the capacitance C and the self-inductance L, both with dimensions of length; a third quantity is the velocity of light c (since we are dealing with an E/M phenomenon). one combination of these three quantities which will produce dimensions of frequency (i.e., one over time) is $\omega = c/\sqrt{LC}$, where again the numerical factor happens to be one. (*Note* In the SI system the formula is $\omega = 1/\sqrt{LC}$).

Determine by dimensional analysis the velocity of propagation υ of a wave in the system of coupled pendulums (see Fig. I(c), p. 126).

By simple inspection of Fig. I(c) we conclude that υ must depend on the lengths l and a, on the mass m and the spring constant κ; we expect to depend also on the acceleration of gravity g and possibly on the wavenumber k. The velocity υ, is equal to ω/k where ω is the common frequency of oscillation of all the masses m. Before we proceed we observe that there are two additive contributions to the potential energy: One is due to gravity and the other to the springs. Since the frequency square is proportional to the kinetic energy which in turn is proportional to the potential energy, we expect that $\omega^2$to be additive, $\omega^2 = \omega_g^2 + \omega_\kappa^2$, where ω_g^2 would be the frequency square, if only the gravity were present and the springs were absent; ω_κ^2 would be the frequency square, in the absence of gravity. We have already shown above that $\omega_g^2 = g/l$. It remains to obtain ω_κ^2 which depends on four quantities: κ, m, a, and k; we observe that κ/m has dimensions of frequency square and the combination ak is dimensionless. Hence the most general expression for ω_κ^2 is of the form: $\omega_\kappa^2 = (\kappa/m)f(ak)$, where f is an arbitrary function of ak, as far as dimensional analysis is concerned. From physical considerations we expect that $f \to 0$ as $k \to 0$, since in this limit all masses move in phase and the springs are inactive. The maximum value of f is obtained when neighboring masses are completely out of phase; this occurs when $2a = \lambda$, or, equivalently, when $ak = \pi$. The function $\sin^2(ak/2)$ satisfies all this requirements for f. Thus, it is not unlikely for someone to guess that f is proportional to $\sin^2(ak/2)$. It turns out that this is actually the case. So

$$\omega^2 = \omega_g^2 + c_1(\kappa/m)\sin^2(ak/2) \tag{5.19}$$

with the numerical constant c_1 being equal to 4. In the absence of gravity and for for $ak \ll 1$, we have $\omega = \sqrt{\kappa/mak} = \sqrt{(\kappa a)/(m/a)}k$; hence, in this case, the velocity of propagation, $\upsilon = \omega/k$, of the wave depends only on the properties of the system and not on the wavelength:

$$\upsilon = \sqrt{B/\rho}, \ B = \kappa a, \ \rho \equiv m/a \tag{5.20}$$

where B is the one-dimensional (1D) analog of the so-called bulk modulus (which is the inverse of the compressibility) and ρ is the 1D mass density.

Determine by dimensional analysis the velocity of propagation υ of sea waves.

This is another example where there are two additive contributions to the potential energy: One is the gravitational potential energy and the other is the potential energy due to surface tension; both reach their minimum equilibrium value when the surface becomes flat. As it was argued before, the additivity of the potential energies implies the additivity of the squares of the frequencies: $\omega^2 = \omega_g^2 + \omega_s^2$. The gravity due ω_g^2 depends for sure on the acceleration of gravity g, possibly on the wavenumber k, and the depth of the sea d; the density ρ does not enter, because its presence would introduce the dimension of mass only on the right side of the equation for ω_g^2. The combination gk has dimensions of inverse time square and the product kd is dimensionless. Hence, the most general expression for ω_g^2 is of the form $\omega_g^2 = gkf_g(kd)$, where the function f_g cannot be determined from the dimensional analysis. The surface tension due ω_s^2 depends, obviously, on the surface tension coefficient σ, possibly on the wavenumber k, on the depth of the sea d, and on the density ρ The combination $\sigma^{\nu_1}k^{\nu_2}\rho^{\nu_3}$ with $\nu_1 = 1, \ \nu_2 = 3, \ \nu_3 = -1$ has dimensions of inverse time square and the product kd is dimensionless. Thus the most general expression for ω_s^2 has the form: $\omega_s^2 = (\sigma k^3/\rho)f_s(kd)$. The functions $f_g(kd)$ and $f_s(kd)$must go to a constant in the limit of very large kd, since, when the depth of the sea is much larger than the wavelength, the depth d plays no role at all; this constant turns out to be equal to one for both f_g and f_s. Thus

$$\upsilon^2 \equiv \frac{\omega^2}{k^2} = \frac{g}{k} + \frac{\sigma k}{\rho}, \quad kd \equiv (2\pi d/\lambda) \gg 1 \tag{5.21}$$

In the opposite limit of $kd \ll 1$, it is expected that ω_s^2 to be negligible in comparison with ω_g^2 and that the relevant length will be only the depth of the sea. For this to happen $f_g(kd)$ must be proportional to kd in this limit; the proportionality constant turns out to be equal to one. So we have

$$\upsilon^2 = gd, \quad kd \equiv (2\pi d/\lambda) \ll 1 \tag{5.22}$$

Equation (5.22) gives the velocity of propagation of a tsunami, which usually has a wavelength much larger than the depth of the sea, d. A detailed calculation shows that $f_g(kd) = f_s(kd) = \tanh(kd)$; this function reproduces the two limits obtained by simple physical considerations. In Fig. 5.1 we plot the velocity υ

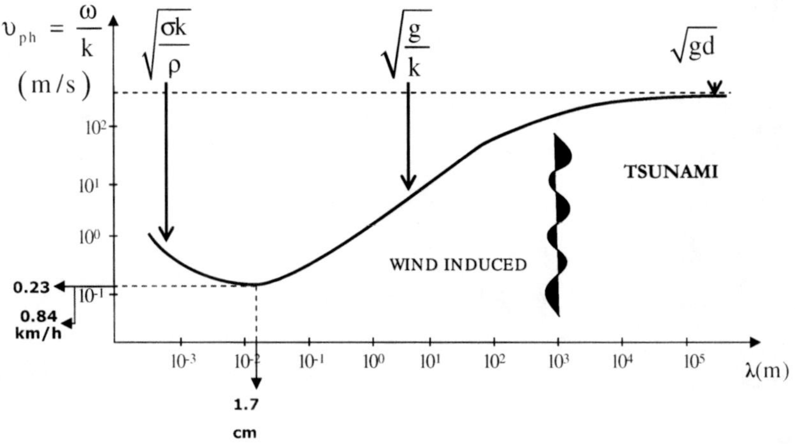

Fig. 5.1 Sea wave velocity versus wavelength. Very short wavelength waves are controlled by surface tension. Very long wavelength waves, as in tsunamis, propagate very fast; e.g. for $d = 2.5$ km, and $\lambda \gg d$, $\upsilon = 158$ m/s $= 569$ km/h. Minimum velocity $\upsilon_{min} = 0.23$ m/s $=$ 0.84 km/h

versus the wavelength λ, taking into account that the surface tension coefficient for water at 293 K is equal to 0.073 J/m^2. Such a plot explains why the wind, as it starts blowing, will set up first sea waves of wavelength of the order of centimeters.

Part II
Structures Mediated by Strong Interactions

Chapter 6
From Quarks and Gluons To Hadrons

Abstract The first step towards composite structures of matter is the combinations of three quarks of total color-charge zero or the combination of a quark and an antiquark of total color-charge zero mainly through the mediation of gluons. Only protons consisting of two up quarks and one down quark are stable. Neutrons are stabilized inside non-radioactive nuclei.

Keywords Hadrons · Baryons · Mesons · Gluons · Photons · Vector bosons

Summary: As it was mentioned in the abstract, three quarks combine to form baryons of total color-charge (c-charge) zero; the only stable baryon, when isolated or as part of non-radioactive nuclei, is the proton consisting of two u quarks and one d quark, all three quarks of different c-charges, so that the total c-charge of the proton to be zero. Neutrons, consisting of two d quarks and one u quark, are metastable, when isolated, while they are stable as constituents of non-radioactive nuclei. All other baryons are metastable. A bound pair of quark/antiquark is what is called a meson, again of total c-charge zero; all mesons are metastable. Baryons and mesons, collectively called hadrons, seem to be the only combinations of quarks in stable or metastable existence. The strong interactions mediated by virtual gluons provide the main mechanism binding together the quarks into baryons and mesons. However, the decay of the latter may involve in general other interactions as well, such as the E/M and the weak ones (see Tables 6.1 and 6.2). Besides protons and neutrons, which make up about 99.97% of the mass around and inside us, a few other metastable mesons and baryons are reaching the Earth from the outer space as cosmic rays or as byproducts of cosmic rays. However, most of the metastable baryons and mesons as well as the metastable leptons and the vector bosons are produced mainly through collisions of particles, such as electrons, protons, positrons, antiprotons, etc. which have been accelerated to reach high kinetic energy by very sophisticated big machines of various types called accelerators or colliders. In the newly operated so-called Large Hadron

E. N. Economou, *A Short Journey from Quarks to the Universe*,
SpringerBriefs in Physics, 1, DOI: 10.1007/978-3-642-20089-2_6,

Collider (LHC) two opposite running beams of protons come to head-on collision; each of these particles will reach a kinetic energy exceeding its rest energy by a factor of about 7,500! With the availability of such a huge energy a variety of numerous particles are produced which are detected by big devices, real "miracles" of science and technology. The enormous amount of collected data are analyzed and interpreted through comparison with the results of the established theoretical scheme, called *the standard model*. The latter is really an admirable intellectual achievement of many years of collective scientific labor. In this chapter, besides a brief introduction to this extensive and expanding field, an estimate of the masses of proton and neutron is offered based on kinetic energy.

6.1 Hadrons and Quantum Chromodynamics: An Outline

The basic elementary process accounting for the formation of baryons and mesons out of quarks is shown schematically in the diagram of Fig. 6.1. Since gluons carry a combination of c-charge/c-anticharge they can also emit or absorb gluons themselves, as shown in Fig. 6.2. By properly combining elementary diagrams, such as those of Figs. 6.1, 6.2 and employing the rules associated with those diagrams, one can in principle calculate theoretically explicit values for experimentally determined quantities, such as the rest mass and the magnetic moment of protons and neutrons, the life-time[1] of an isolated neutron, etc. Before attempting to calculate such specific quantities, one has to answer first the following general questions:

a. Why do the bound combinations of quarks, stable or metastable, have to be colorless (i.e., of zero total c-charge)?
b. Why do only bound combinations of three quarks or bound quark/antiquark pairs exist? In other words, why stable (or metastable) zero total c-charge bound combinations of $3n$ quarks, or stable (or metastable) bound combinations of n quarks/n antiquarks do not exist?
c. Could real gluons (as opposed to virtual ones) be constituents of stable (or metastable) bound clusters?

We know that the answers to such questions can be obtained by the general idea already presented in Chap. 4: We have to calculate the total energy of the observed composite particles and show that it is considerably lower than the energy of the other hypothetical particles mentioned above. The implementation of this obvious approach meets serious practical difficulties due to the high strength a_s of the

[1] There are two related times which characterize the rate of exponential decay of metastable structures. If initially ($t = 0$) there were N_o such structures, their number at a later time t is $N(t) = N_o \exp(-t/\tau)$, where τ is called their *mean lifetime*. We define also the *half-life*, $t_{1/2}$, as the time at which the remaining number $N(t_{1/2})$is half the initial number N_o; obviously we have $\exp(-t_{1/2}/\tau) = 1/2$, or $t_{1/2} = \tau \ln 2 = 0.693\tau$

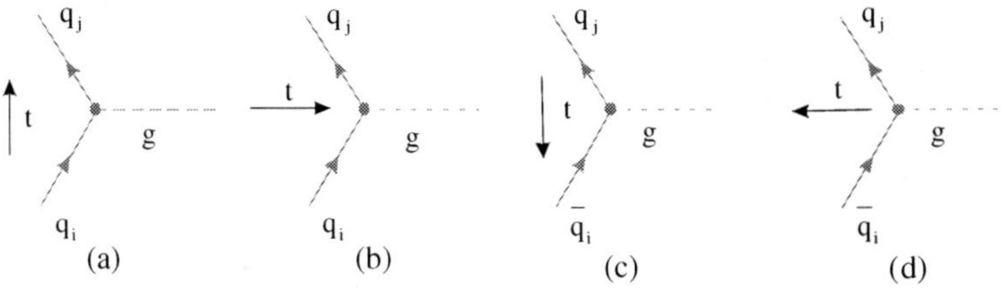

Fig. 6.1 The same basic elementary diagram represents different basic elementary processes depending on its different orientation relative to the arrow of time: **a** a quark q of given c-charge emits or absorbs a gluon remaining a quark in the same family and with the same e-charge; the c-charge may change, if the gluon is not colorless (e.g., if $q_i \equiv u_G$ and if it emits a $g = g_{G\bar{B}}$, then $q_j = u_B$); **b** a pair of quark/antiquark of the same family and of opposite e-charge is annihilated and produces a gluon; if the c-charges of the pair are, e.g. $R, \overline{G}$, then the gluon is $g_{R\overline{G}}$; **c** an antiquark $\overline{q}_j$ emits or absorbs a gluon, remaining an antiquark $\overline{q}_i$ of the same family and the same e-charge; the c-anticharge of the antiquark $\overline{q}_i$, may be different than that of $\overline{q}_j$, if the gluon is not colorless; **d** a gluon is annihilated by creating a pair of quark/antiquark of the same family; e-charge and c-charge are conserved in this process as in all processes

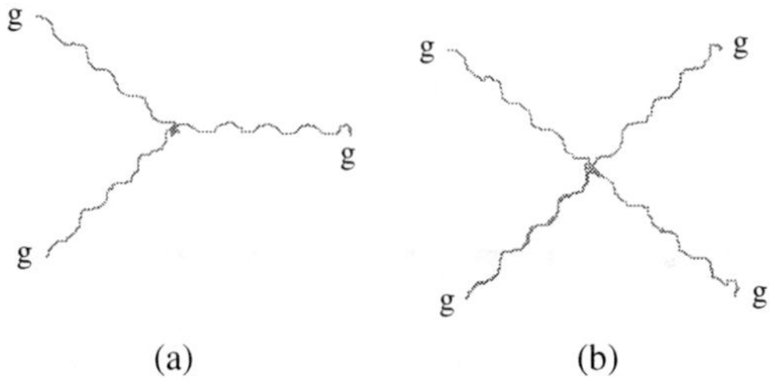

Fig. 6.2 Elementary diagrams representing elementary processes due to strong interactions and involving only gluons: **a** direct interaction among three gluons, **b** direct interaction among four gluons

Table 6.1 Some of the baryons formed out of three quarks of total c-charge zero. Only protons and neutrons are stable. (Neutrons are stable only inside stable nuclei) [6.1]. See also Pr. 6.1

Name/ symbol	Composition	Mass/m_e	Size (fm)	Electric charge ($\|e\|$)	Mean lifetime	Main process of decay
proton/p	uud	1,836,15	0,84	1	$>10^{32}$yr	–
neutron/n	udd	1,838,68	0,84	0	889	$\rightarrow pe\overline{\nu}_e$
Δ^{++}	uuu	2,411	–	2	5.5×10^{-24}	$\rightarrow p\pi^+$
Λ^o	uds	2,183		0	2.63×10^{-10}	$p\pi^-$, $n\pi^o$
Σ^o	uds	2,334		0	6×10^{-20}	$\rightarrow \Lambda^o\gamma$
Σ^+	uus	2,318		1	0.8×10^{-10}	$\rightarrow p\pi^o$, $n\pi^+$
Σ^-	dds	2,343		−1	1.48×10^{-10}	$\rightarrow n\pi^-$
Ξ^o	uss	2,573		0	2.9×10^{-10}	$\rightarrow \Lambda^o \pi^o$
Ξ^-	dss	2,586		−1	1.64×10^{-10}	$\rightarrow \Lambda^o\pi^-$
Ω^-	sss	3,273		−1	0.82×10^{-10}	$\Lambda^o K^-$, $\Xi^o\pi^-$
Λ_c	udc	4,471		0	2.1×10^{-13}	$pK^-\pi^+$

Table 6.2 Some of the mesons formed out of a quark and an antiquark; all of them are metastable [6.1]. See also Pr. 6.2

Symbol	Composition	Mass $/m_e$	Electrical charge ($\|e\|$)	Spin	Mean lifetime(s)	Main process of decay
π^o	$\frac{1}{\sqrt{2}}(u\bar{u} - d\bar{d})$	264	0	0	0.84×10^{-16}	$\gamma\gamma$
π^+	$u\bar{d}$	273	1	0	2.603×10^{-8}	$\mu^+ \nu_\mu$
π^-	$d\bar{u}$	273	−1	0	2.603×10^{-8}	$\mu^- \bar{\nu}_\mu$
K^o	$d\bar{s}$		0	0	K_L^o 5.17×10^{-8}	$\pi^o \pi^o \pi^o$
$\overline{K}^o$	$s\bar{d}$	970	0	0	K_s^o 0.892×10^{-10}	$\pi^+ \pi^-$
K^+	$u\bar{s}$		1	0	1.237×10^{-8}	$\mu^+ \nu_\mu$
K^-	$s\bar{u}$		−1	0	1.237×10^{-8}	$\mu^- \bar{\nu}_\mu$
J/ψ	$c\bar{c}$	6070	0	1	7.2×10^{-21}	Mainly hadrons $e^+e^-, \mu^+\mu^-, \eta(\pi^+\pi^-)\pi^o$
D^o	$c\bar{u}$		0	0	4.15×10^{-13}	$K^-\pi^+\pi^o$ κ.α.
$\overline{D}_o$	$u\bar{c}$		0	0		
D^+	$c\bar{d}$	3650	1	0	1.04×10^{-12}	
D^-	$d\bar{c}$		−1	0		
B^+	$u\bar{b}$		+1		1.62×10^{-12}	
B^-	$b\bar{u}$		−1			
B_d^o	$d\bar{b}$		0			
$\overline{B}_d^o$	$b\bar{d}$	10315	0		1.56×10^{-12}	
B_s^o	$s\bar{b}$		0			
$\overline{B}_s^o$	$b\bar{s}$		0			

strong interactions. As a result of a_s being of the order of unity, each of a huge number of diagrams of ever increasing number of vertices and complexity contributes to the total energy an amount of the same order of magnitude. This makes the diagram-based explicit calculations practically impossible [6.1]. To understand this difficulty, it may help to compare the theory of electron–photon E/M interaction (named quantum electrodynamics, QED) with the theory of quark-gluon strong interaction (named quantum chromodynamics, QCD [6.2]). In QED there are no diagrams similar to those shown in Fig. 6.2, since the photons do not have e-charge and, hence do not interact directly with each other; as a result, the diagrams in QED are much simpler. Moreover, the coupling constant a in QED is only 1/137 (see Table 2.2), in contrast to $a_s \approx 1$ in QCD; as a result, and in contrast to QCD, there is no need in QED to keep diagrams with more than a few vertices.[2] The contributions of a few simple diagrams in QED suffice to give theoretical results in impressive agreement with the experimental data. In conclusion, the strength of the coupling constant $a_s \approx 1$ and the extra complication introduced by the elementary diagrams of Fig. 6.2 prevent us from

[2] Each additional vertex in a diagram introduces a factor of $a = 1/137$ in QED, while the corresponding factor in QCD is a_s, i.e., approximately equal to one.

repeating in QCD the phenomenally successful calculations of QED. Nevertheless, a special very laborious computational scheme has been developed which predicts the masses of several baryons and mesons in reasonable agreement with the corresponding experimental data [6.3]. Again the analogy with QED is helpful: We mentioned before (p. 47) that the range r_o of each interaction is related to the mass m of the corresponding ic-particle through the formula $r_o = \hbar/mc$. That is why the range of Coulomb interaction energy between two e-charged particles mediated by photon exchange is infinite. By analogy we should expect that the range of the strong interaction energy between two quarks mediated by gluons to be infinite. However, we know experimentally that this is not true: The range of the strong interactions is quite small of the order of 10^{-15} m. How does QCD account for this apparent discrepancy? There are strong indications, although no rigorous proof, within the framework of QCD that the interaction energy between two quarks (or between two c-charged particles) tends to infinity as their separation tends to infinity. Thus, the energy cost to completely isolate a quark or a c-charged particle is infinite. In other words, it is forbidden by energy considerations to have isolated quarks, or other entities carrying a non-zero c-charge. This argument provides an answer to question (a) above and at the same time explains why the range of the strong forces is so short: Quarks or any other c-charged particles are always locked-up within c-neutral structures such as protons, neutrons, pions, etc. This *confinement* explains automatically the effectively short-range nature of the strong interaction. What is going to happen, if we try to extract a quark from a c-neutral structure by supplying a lot of energy? The system may be deformed and stretched in complicated ways until enough energy is provided to one of the virtual gluons keeping the quarks together; then the gluon may break to a quark/antiquark pair as shown in Fig. 6.1d. The end result will to extract a meson, i.e. a c-neutral structure and leave behind another c-neutral structure containing the same number of quarks as the initial one. A similar mechanism is at work when two c-neutral composite particles (e.g., a neutron and a proton) come close together; their interaction, instead of being described through gluon exchange, may be described in a simpler, but essentially equivalent way, as the creation and exchange of virtual pions and possibly other mesons between the two c-neutral particles. This meson mediated description of the interaction of proton/proton, proton/neutron, and neutron/neutron will be used in the next chapter to account for the formation of atomic nuclei out of protons and neutrons. This meson mediated description may break down when the two c-neutral composite particles are forced to come so closed together as to almost merge into one. In this case the more fundamental description in terms of quarks and gluons may be necessary In any case it is shown experimentally that the merging of the pair proton/proton (or the pair proton/neutron, or the neutron/neutron) into a single six-quark composite particle is not energetically favorable and does not occur spontaneously. On the other hand, the possibility of a large number of quarks and gluons merging together to form a dense quark/gluon "soup", called *quark/gluon plasma*, under conditions of high temperature and concentration, has been studied extensively.

It must be pointed out that the theoretical scheme of QCD, as summarized by the elementary diagrams shown in Figs. 6.1 and 6.2, shows that the coupling "constant" a_s is not so constant: It does depend on the energy or, equivalently, on the distance. As the energy increases or the distance decreases the coupling "constant" is reduced and tends to a very small value. This is the feature of *asymptotic freedom* [6.3]. It means that the theory of QCD at say 10^9 eV with the corresponding value of the coupling "constant" a_s is equivalent to the very same theory at 10^{10} eV but with a smaller value of a_s. This length (or energy) dependence of the QCD coupling "constant" is reminiscent of the electrostatic screening of the interaction between two electrons in a metal by the dielectric "constant" due to all the other electrons. The inverse of the dielectric "constant" which is the analog of the coupling "constant" in QCD takes its largest value (=1) as the distance of the two electrons becomes very small, and it tends to zero as this distance becomes very large. In QCD, a_s behaves in the inverse manner: The coupling "constant", a_s, takes larger and larger values as the distance increases and, hence, it prevents the separation of a c-charged entity; and it takes very low values for short distances (high energies). Another difference between the electrostatic screening and the QCD asymptotic freedom is that in the latter the length dependence of the coupling "constant" is taking place in the vacuum as a result quantum vacuum fluctuations.

We shall conclude this chapter by attempting to estimate the rest energy of the proton (or the neutron) by assuming asymptotic freedom at its extreme, i.e., zero coupling constant for the quarks inside the proton (or the neutron). Then, the rest energy equals the rest energy of the three quarks plus the kinetic energy of the confinement within the volume of the proton. The center of mass is free to move everywhere; so its three degrees of freedom (corresponding to a single particle in three dimensions) are not subject to confinement. The latter concerns only the relative motion of the quarks i.e., the remaining six degrees of freedom (approximately corresponding to two particles in three dimensions). Thus, we can write for the average energy of a proton whose the center of mass is at rest:

$$\langle \varepsilon \rangle \approx 2\left\langle \sqrt{m^2c^4 + c^2p^2} \right\rangle \approx 2\langle cp \rangle \tag{6.1}$$

where m is an appropriate linear combinations of the rest masses, m_1, m_2, m_3 of the quarks. Notice that the kinetic energy of confinement is in the extreme relativistic limit, $cp \gg mc^2$, since $\langle \varepsilon \rangle$ is of the order of 1,000 MeV, while mc^2 is less than 10 MeV. The quantity $\langle cp \rangle$ has been estimated in Chap. 3, Eq. 3.7: $\langle cp \rangle \approx 3\hbar c/V^{1/3}$. In the present case the radius of the proton (or neutron) is 0.8 fm $= 1.5 \times 10^{-5}$ atomic units and the volume to the 1/3 power is 2.42×10^{-5} atomic units; then the numerical value of $\langle cp \rangle$ is 462 MeV. Ignoring the Coulomb repulsion of the two u quarks in the proton we obtain for the proton and neutron mass:

$$m_pc^2 \approx m_nc^2 \approx 928.3\,\text{MeV} \tag{6.2}$$

surprisingly close to the experimental values of 938.27 and 939.57 MeV, respectively.

Chapter 7
From Protons and Neutrons to Nuclei

Abstract Protons and neutrons combine to form stable and metastable nuclei. The total energy of the nuclei depends on the number of protons and neutrons involved. Having this dependence of the total energy allows us to account for many properties of the various nuclei.

Keywords Binding energy · Fission · Fusion · α-decay · β-decay · Fissionable isotope

Summary The energy E of a nucleus consisting of Z protons and N neutrons ($Z + N \equiv A$) is equal to $Zm_pc^2 + Nm_nc^2 - B$ where B is the binding energy of all the A nucleons. The formula for $-B$ is the following:

$$-B = -\alpha A + \beta A^{2/3} + \gamma \frac{Z(Z-1)}{A^{1/3}} + \alpha_K \frac{(A-2Z)^2}{A} + \delta; \quad A \equiv Z + N \tag{7.1}$$

where $\delta = \mp 34/A^{3/4}$ in MeV the upper sign holds if both N and Z are even (a situation to be denoted as e,e) and the lower one if both N and Z are odd (a situation to be denoted as o, o). If A is odd, then $\delta = 0$. The radius R of a nucleus is given by $R = c_s A^{1/3}$, where the "constant" c_s is between 1.18 (for nuclei of large A) and 1.3 (for nuclei of small A) with a typical value $c_s \approx 1.24$. Having an expression of the total energy E as a function of Z and $A = Z + N$, we minimize it with respect to Z keeping A constant; we obtain thus the percentage of protons (and hence of neutrons) in a nucleus of A nucleons[1]; the result is

$$\frac{Z}{A} = \frac{1 + \frac{(m_n - m_p)c^2}{4\alpha_K} + \frac{\gamma A^{-1/3}}{4\alpha_K}}{2 + \frac{\gamma A^{2/3}}{2\alpha_K}} \tag{7.2}$$

[1] *Nucleon* is the common name used for both the proton and the neutron.

E. N. Economou, *A Short Journey from Quarks to the Universe*, SpringerBriefs in Physics, 1, DOI: 10.1007/978-3-642-20089-2_7,

In this chapter we derive (7.1) by employing an additional simplifying approximation for the residual strong interaction between pairs of nucleons. Then, having (7.1) and (7.2), we are in a position to study the stability of various nuclei (i.e., to predict how many and which stable nuclei exist, how many and which metastable (i.e., radioactive) nuclei exist, and which combinations of Z protons and N neutrons are unstable and, hence, do not exist). We can also provide quantitative answers to various questions such as the following: How can one extract energy from nuclei? Why are the byproducts of a nuclear reactor radioactive? Why out of all elements is Uranium of such critical importance in the extraction of nuclear energy? Why is Uranium-235 fissionable, while Uranium-238 is not? Etc.

7.1 Calculating the Total Energy

We already mentioned in Chap. 6 (p. 49) that there is an attractive residual strong interaction between two neighboring nucleons; this attraction appears when their distance[2] d is in the range of about 1 fm to about 4 fm. For smaller distances the interaction becomes repulsive as a result of the fact that the merging of two nucleons into a single six-quark particle is energetically unfavorable. For distances larger than 4 fm the residual interaction soon becomes negligible, since beyond 4 fm it decays exponentially as $\exp(-d/d_o)/(d/d_o)$, where $d_o = \hbar/cm_\pi$ is related to the mass m_π of the pions (p. 48). On the basis of these remarks we expect that the mean distance between neighboring nucleons in a nucleus to be somewhere between 1 and 4 fm but closer[3] to the 1 fm. Actually this mean distance is about 2 fm, which means that the percentage of the volume of the nucleus occupied by the nucleons is on the average equal to $(4\pi/3)r_o^3/2^3 = 0.31$, where $r_o = 0.84$ fm is the radius of each nucleon. If we include the non-available empty space among randomly placed touching spheres of equal size this 0.31 is increased by about 60% to become 0.50. Thus the volume V' of the really available empty space for the motion of the nucleons is 50% of the volume V of the nucleus. These numbers could be obtained theoretically, if the total energy of each nucleus were expressed in terms of the mean distance d between neighboring nucleons. However, the attractive residual strong interaction between neighboring nucleons is a very complicated function not only of d but also of the relative orientation of their spins, of their relative motion, etc. As a result, in what follows we will accept the empirical value of $d = 2$ fm and the empirical *mean* value of the attractive interaction $V_s \approx -11.4$ MeV corresponding to $d = 2$ fm. We are now in a position to estimate the total energy E of a nucleus of Z protons and N neutrons. There are three contributions to E:

---

[2] d is the distance between the centers of the two nucleons.

[3] Being closer to 1 fm, it fully explores the attraction but it increases the quantum kinetic energy.

a. The residual attractive strong interaction gives a contribution E_s equal to $N_{\text{pairs}}V_s$ where N_{pairs} is the number of interacting pairs of nucleons. Because of the short range character of the residual strong interaction only nearest neighbor pairs interact. Thus

$$N_{\text{pairs}} = \frac{1}{2}A_B N_B + \frac{1}{2}A_S N_S$$

where A_B and A_S are the number of nucleons in the bulk (i.e., in the interior of the nucleus) and in the surface, respectively. $A_S \sim R^2 \sim A^{2/3}$ and $A_B = A - A_S$ where $R \sim A^{1/3}$is the radius of the nucleus. $N_B \approx 8$ is the average number of nearest neighbors for a nucleon in the bulk and $N_S \approx 5$ for a nucleon at the surface. A rough estimate of A_S gives the following result: $A_S \approx A^{2/3}$. Hence the residual strong interaction gives:

$$E_s = -45.6A + 17.1A^{2/3} \text{ in MeV} \tag{7.3}$$

b. The second contribution to $-B$ is due to the kinetic energy of the nucleons; it is of repulsive nature, and it is given (according to (3.9) p. 21) by

$$E_K = 2.87\frac{\hbar^2}{mV'^{2/3}}\left(Z^{5/3} + N^{5/3}\right)$$

Taking into account that $V' = 0.5V = (0.5)4\pi R^3/3$, $R \approx 1.24A^{1/3}$ (in fm) and

$$Z^{5/3} + N^{5/3} \approx \frac{1}{2^{2/3}}\left[A^{5/3} + \frac{5}{9}\frac{(N-Z)^2}{A^{1/3}}\right]$$

we can rewrite E_K as follows

$$E_K \approx 29.75\,A + 16.5\frac{(N-Z)^2}{A} \quad \text{in MeV} \tag{7.4}$$

c. The Coulomb repulsion among protons is

$$E_c = \frac{1}{2}\sum\nolimits_{i,j=1}^{Z}\frac{e^2}{r_{ij}} = \frac{1}{2}\frac{Z(Z-1)e^2}{r},\ r \simeq \frac{5}{6}R \text{ or } E_C \approx 0.71\frac{Z(Z-1)}{A^{1/3}}\text{MeV} \tag{7.5}$$

A final contribution to the total energy of a nucleus has to do with the discrete nature of the energy levels that a nucleon can occupy. Each such level, in view of Pauli's exclusion principle, can be occupied by up to four nucleons (two protons with opposite spin and two neutrons with opposite spin). Thus in an even,even nucleus, under conditions of minimum total energy, the highest occupied level will be occupied by exactly two protons and two neutrons. Let us now change one of the neutrons to a proton transforming the nucleus to an odd, odd one; then one of the resulting three protons has to move to the next higher level, increasing thus the total energy by $2\delta \equiv \varepsilon_{n+1} - \varepsilon_n$, where ε_n, ε_{n+1} are the highest occupied levels for the e,e and the o,o case, respectively. In the case of odd A the highest occupied level will have, let us say, two neutrons and one proton. Changing one neutron to proton will leave the highest occupied level still having three nucleons; thus no reoccupation of the levels will take place and, hence, no change in energy is expected due to such a process. These arguments account for the appearance of the last term in (7.1). Adding together (7.3), (7.4), and (7.5) and including the δ term we end up with (7.1); the values we obtained for the coefficient in (7.1) are as follows:

$$\alpha \approx 45.6 - 29.75 = 15.85,\ \beta \approx 17.1,\ \gamma \approx 0.71,\ \text{and}\ \alpha_K \approx 16.5\ \text{all in MeV} \tag{7.6}$$

while by adjusting these coefficients as to best fit the experimental data we have [7.1]:

$$\alpha = 15.75,\ \beta \approx 17.8,\ \gamma \approx 0.71,\ \text{and}\ \alpha_K \approx 23.69\ \text{all in MeV} \tag{7.7}$$

Our simplistic approach produced results unexpectedly close to the best fit data (with the exception of α_K). By differentiating (7.1) with respect to Z under constant A, taking into account that $N = A - Z$, and replacing the values of the coefficients as given by (7.7), we obtain (7.2), or more explicitly:

$$\frac{Z}{A} = \frac{1.014 + 0.0075A^{-1/3}}{2 + 0.015A^{2/3}} \tag{7.8}$$

Notice that weak interactions allow a nucleus to change neutrons to protons and vice versa until the optimum ratio Z/A, as given by (7.8), is achieved. (see Figs. 2.3 and 2.5 and recall that a proton consists of u,u,d quarks, while a neutron of u,d,d quarks)

It is clear from (7.1) that the quantum kinetic energy becomes minimum, when $Z = N = A/2$, while the Coulomb energy becomes minimum, when $Z = 0$ or $Z = 1$; also protons are slightly favored because they have a smaller rest energy. All these conflicting requirements, each according to its "weight", determine the optimum ratio Z/A which minimizes the total energy, as shown in (7.2). Indeed the term $(m_n - m_p)c^2/4\alpha_K \approx 0.014$ is there because of the difference in the rest masses and it slightly increases the percentage of protons. The other terms on the right hand side (r.h.s.) of (7.2) are proportional to γ/a_K and are due to the Coulomb repulsion among protons normalized by the kinetic energy coefficient a_K. If this repulsion were absent, the ratio Z/A would be a constant equal to 0.507. In the presence of the Coulomb repulsion this ratio decreases monotonically with increasing A starting with a value about equal to 0.50 for small A and reaching about 0.39 for large A, such as $A = 238$ corresponding to Uranium. This behavior of Z/A is due to the fact that the "weight" of the Coulomb contribution relative to that of the quantum kinetic energy is almost negligible for small A, but it is

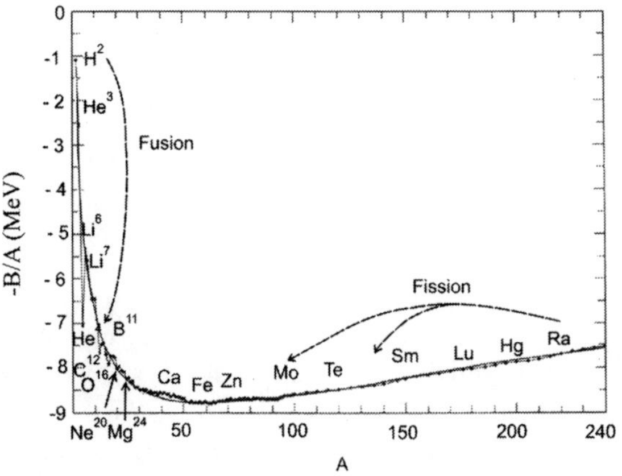

Fig. 7.1 The energy reduction per nucleon, $-B/A$, (as a result of the nucleus formation) versus the number of nucleons A. The continuous line is based on (7.1) without the δ term and on (7.2), (7.7). The dots are the experimental data. The discrepancies between the continuous line and the experimental data almost disappear with the inclusion of the δ term. (Check the extreme case of He^4). The rise of the curve for small A is due to the increased number of nucleons at the surface where the exploitation of strong nuclear force is less effective. The rise for large A is due to the repulsive Coulomb forces. This rise makes possible the extraction of energy by *fission* (i.e., by splitting of an initial large nucleus to two smaller ones of comparable size); it is also responsible for the non-existence of nuclei with A larger than about 240 (because, then, the emission of a He^4 nucleus is spontaneous). Energy can be extracted also by *fusion*, i.e., by merging together two small nuclei into a single one, as shown in the figure; this process takes place in stars and in the thermonuclear bombs

increasing faster with A than that of the kinetic energy[4] and, as a result, the balance tips to more neutrons than protons. Substituting (7.8) and (7.7) in (7.1) we obtain $-B$ as a function of A. In Fig. 7.1 we plot the ratio $-B/A$ versus A. It is preferable to plot $-B/A$ rather than $+B/A$, as several books do, because the former gives essentially the total energy per nucleon, without including the almost constant quantity $(Zm_pc^2 + Nm_nc^2)/A$.

7.2 Questions and Answers

The formula for the total energy of a nucleus allows us to provide answers to some reasonable questions:

1. *Why is the nucleus such an enormous giant in energy*?

[4] The kinetic energy term increases linearly with A, while the Coulomb term increases almost proportionally to $A^{5/3}$ (see (7.4) and (7.5)).

Because it is such a minute dwarf in size. Its energy scale is determined by the quantum kinetic energy given by (3.6), $\varepsilon_K \approx \hbar^2/m_p r_o^2 \approx 40$ MeV for $r_o \approx 1$ fm.

2. *Why are not there nuclei of A larger than about 240?*

Because it is energetically favorable and feasible such a nucleus to break into two pieces. Let us examine the case where one of the two pieces is a He^4 nucleus ($Z = N = 2$). For this process, known as α-decay, to happen we must have the inequality $E(A) \geq E(A - 4) + E(4)$ or $B(A) \leq B(A - 4) + B(4)$. As shown in Fig. 7.1, $B(4) \approx 4 \times 7.074$ and the $-B/A$ can be approximated for large $A(A > 100)$ by a straight line, $-B/A \approx -9.45 + 0.008A$. Hence the inequality becomes $9.45A - 0.008A^2 \leq 9.45(A - 4) - 0.008(A - 4)^2 + 4 \times 7.074$ or $A \geq 150.5$

However, the requirement that the rest energy of the final state to be lower than that of the initial is not enough for the reaction to proceed; an energy barrier may stand between the initial and the final state preventing the realization of the process. If the barrier is not very high and wide and the reduced mass of the two fragments is not so large, quantum mechanical tunneling may allow the penetration of the barrier to take place; otherwise, an external supply of energy may be needed in order to take the initial system over the barrier. Anyway, our previous theoretical result shows that the α-decay would not take place in nuclei with A smaller than about 150, while it may take place for A larger than 150. The experimental data are in good agreement with these predictions: No nucleus with $A<144$ exhibits α-decay; between $A = 144$ and $A = 151$ there only five nuclei undergoing α-decay, but with extremely long lifetimes (of the order of 10^{16}years!). For A between 152 and 208 there are a few nuclei undergoing α-decay and many others which are stable. For $A > 208$ all nuclei are metastable. E.g., U^{238} through a series of α-decays (and β-decays) ends up as Pb^{206}.

Let examine also the case where a nucleus may break spontaneously into two fragments of comparable size. For simplicity we shall take the two fragments to be equal, although unequal fragments would produce smaller reduced mass and, hence, higher probability for tunneling. Using (7.1) and (7.7) we find that the rest energy difference between the initial and the final state equals to $\Delta E \approx 0.2635\, Z^2/A^{1/3} - 4.6266\, A^{2/3}$which is positive, if $(Z^2/A) \geq 17.5$; however, since the reduced mass of the fragments in this case is quite large, $Am_p/2$, it is expected that the potential barrier, if any, would be impenetrable. Thus we have to calculate the height of the energy barrier, E_B, and find when it would become zero. As it is shown in Fig. 7.2, $E_B = E_C - \Delta E$, where E_C is the maximum value of the Coulomb repulsion as the two fragments are brought back towards the initial state, but before the residual strong interaction between them would become appreciable: $E_C \approx (Z_1 Z_2 e^2/d_i)$, where $d_i \approx R_1 + R_2 + 3.5\,\text{fm} \approx 1.57(A_1^{1/3} + A_2^{1/3})$ for $A \approx 300$. In the present case $Z_1 = Z_2 = Z/2$ and $A_1 = A_2 = A/2$. Finally, we obtain for E_C, $E_C \approx 0.1473 Z^2 A^{-1/3}$. Combining this value of E_C with ΔE as obtained above, we have for E_B:$E_B = 4.6266A^{2/3} - 0.1162Z^2/A^{1/3}$ which becomes zero when $Z^2/A \approx 40$; this implies, in view of (7.8), that for $A \approx 300$, $Z \approx 113$. Already,

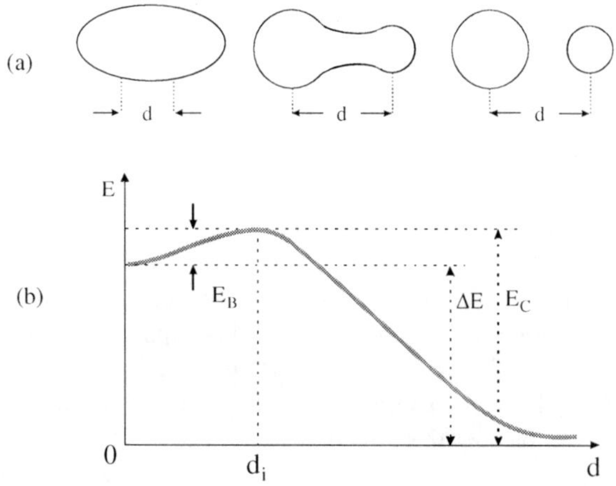

Fig. 7.2 **a** Fission of a large nucleus, such as U^{235}, proceeds usually through the absorption of an external neutron accompanied by the elongation of the nucleus; next a neck is created and finally the nucleus breaks into two fragments and 2–3 isolated neutrons; these neutrons may cause further fissions, leading thus to the so-called nuclear chain reaction. **b** The continuous curve gives the total energy as a function of the separation distance d; the zero of energy has been chosen to be the sum of the rest energies of the two fragments as $d \to \infty$ but before any β-decay occurs. ΔE is the energy difference between the initial and the final rest energies; E_C is the maximum energy calculated by imaginarily reversing the process of fission and using Coulomb repulsion up to a separation distance d_i just before the strong interactions become appreciable; $E_B = E_C - \Delta E$ is the height of the energy barrier

metastable short-lived nuclei of $Z = 116, A = 292$ and $Z = 114, A = 298$ have been made artificially.

It is worthwhile to calculate the quantities E_C and ΔE for the important case of U^{235}, using the above formulae: We obtain $\Delta E = 182.2$ MeV, while the corresponding experimental value is 180 MeV, and $E_C = 193.94$ MeV, while the corresponding experimental value is 186 MeV. Thus the experimental value for the height of the energy barrier is 6 MeV.

3. *Why are the fission fragments of* U^{238} *or* U^{235} *undergoing β-decay?*

In the fission fragments the percentage of neutrons is almost[5] the same as in the parent nucleus, i.e., about 60.3%. Let us assume that the smaller fragment has $A_1 = 106$ and the larger has $A_2 = 130$. The equilibrium percentage of neutrons for the smaller and the larger fragments are, according to (7.8), equal to 56.5 and 57.5%, respectively. Thus both fragments have more neutrons than the equilibrium number corresponding to their size. As a result, one or two neutrons in each

[5] During the fission 2–3 isolated neutrons are released.

fragment will change to protons by emitting an electron and an antineutrino (see Fig. 2.3 p. 14), i.e., a β-decay will take place.[6]

4. *How many stable nuclei do exist?*

Since for each A there is one optimum Z which minimizes the total energy, it is reasonable to expect that we must have as many stable nuclei as the maximum number of A corresponding to a stable nucleus. The only exception to this conclusion may occur, if the optimum Z is exactly in the middle between two successive integers; in this case either of these two integers would give the same total energy. Such an improbable coincidence almost appears for four odd nuclei: $A = 87$ and $Z = 38$ or 37; $A = 113$ and $Z = 49$ or 48; $A = 115$ and $Z = 50$ or 49; $A = 123$ and $Z = 51$ or 52. The above argument does not apply to nuclei with even A. To see why consider three such nuclei of the same even A: (Z, N), $(Z-1, N+1)$, $(Z-2, N+2)$ with Z being even. Let us assume that the nucleus (Z, N) has the lowest energy. Because of the δ term in the total energy, it is quite possible that the o,o nucleus $(Z-1, N+1)$ to have higher energy than any of the other two; then the e,e nucleus $(Z-2, N+2)$, although of higher energy than the (Z, N), is stable, because to end up in the latter, it has to go through the intermediate step of becoming a $(Z-1, N+1)$ nucleus by a β-decay; but this β-decay is not allowed, since the $(Z-1, N+1)$ nucleus has higher energy than the $(Z-2, N+2)$ one [7.1, 7.2]. Thus for the same even A there may be two or even three stable e,e nuclei; as a result there are in total about 270 stable nuclei corresponding to values of A from 1 to 208 and values of Z from 1 to 82 excluding the $Z = 43$ (Tc) and $Z = 61$ (Pm) which are artificial radioactive nuclei.

5. *Why is U^{235} fissionable, while U^{238} is not?*

Spontaneous fission of a Uranium nucleus is a very rare event: There is only one spontaneous fission for every 50 millions of α-decays. Neutron assisted fission is quite another story: Indeed, if an external neutron is incorporated in a Uranium nucleus, let us say (A, Z), a new Uranium nucleus (A + 1, Z), will result; the energy of this new nucleus (excluding the rest energy of the nucleons) is the ground state energy plus an amount equal to $\varepsilon_K + \varepsilon_n^*$, where ε_K is the kinetic energy of the neutron before its incorporation in the initial nucleus (A, Z) and ε_n^* is the energy increase above the ground state energy, $E(A+1, Z)$, of the nucleus $(A+1, Z)$ due to the incorporation of a neutron of zero kinetic energy; conservation of energy requires that $E(A+1, Z) + \varepsilon_n^*$ to be equal to the ground state energy $E(A, Z)$ of the nucleus (A, Z) (again the rest energy of the nucleons is not included). Hence, $\varepsilon_n^* = E(A, Z) - E(A+1, Z) = B(A+1, Z) - B(A, Z)$. Let us calculate first ε_n^* for the nucleus U^{235} for which $A = 235$. We have mentioned before that $B(A) \approx 9.45A - 0.008A^2 + \delta$, where the $+\delta$ appears because the nucleus (A + 1, Z), in contrast to (A, Z), is even, even. Thus we have for $\varepsilon_n^*(235)$:

[6] The nucleus resulting after a β-decay is usually in an excited state; it returns to its ground state by emitting a high energy photon in the γ range; this process is known as *γ-radioactivity*.

$\varepsilon_n^*(235) \approx 9.45 - 0.016A + \delta \approx 6.25$ MeV. We conclude that the energy, 6.25 MeV, supplied to U^{235} by the incorporation of a neutron of zero kinetic energy is higher than the height of 6 MeV of the energy barrier. Therefore a neutron of zero kinetic energy incorporated to the nucleus U^{235} will lead to the immediate fission of the latter. Let us repeat now the calculation for U^{238}: $\varepsilon_n^*(238) = B(239) - B(238) \approx 9.45 - 0.016A$ - $\delta \approx 5.1$ MeV. In the case of U^{238} the + δ term is associated with the e,e nucleus (238,Z) and appears with negative sign in ε_n^*. Since $\varepsilon_n^*(238) \approx 5.1$ MeV is less than the height of the energy barrier (6 MeV), we conclude that U^{238} is not *fissionable*, i.e., the incorporation of a neutron of zero kinetic energy does not lead to its immediate fission. According to our estimates the neutron must have a kinetic energy of at least 0.9 MeV in order to lead to the immediate fission of U^{238} (by being incorporated to the latter); the corresponding experimental value is 1.1 MeV. Besides U^{235}, which is the only naturally occurring fissionable isotope, fissionable nuclei are the plutonium −239 (Pu^{239}) and the U^{233}; both of them are produced artificially by neutron incorporation to the non-fissionable nuclei U^{238} and Th^{232} respectively, and by subsequent β-decays.

Part III
The Realm of Electromagnetism

The structure of matter from the level of atoms up to the level of an asteroid, which has linear dimension less than a few hundred kilometers, is dominated by only one of the four interactions: The electromagnetic one whose strength is determined by the charge e (of the proton). The quantum kinetic energy, which counterbalances this force depends (according to Eq. 2.6) on the ratio $\hbar^2/m$. The dominant kinetic energy will be the one with the smaller mass in the denominator, i.e., the mass of the electron, m_e.

The three quantities e, $\hbar$, m_e define a system of units, called atomic system of units (asu). In this asu, the units of mass, length, and time are the following:

unit of mass: m_e, the rest mass of electron $\approx 9.109 \times 10^{-31}$ kg
unit of length: Bohr radius, $a_B \equiv \hbar^2/m_e\, e^2 \approx 0.529$ A
unit of time: $t_0 = \hbar^3/m_e\, e^4 \approx 2.42 \times 10^{-17}$ s

The units of various other physical quantities in the asu are given in Table 2 (p. 142). In many cases, we shall find it more convenient to use the triad a_B, $\hbar$, m_e, rather than e, $\hbar$, m_e.

Chapter 8
From Nuclei and Electrons to Atoms

Abstract In this chapter the main properties of the electrons trapped around each nucleus to form atoms are presented. Of special interest is the size of the atoms as determined by the outer electrons as well as the motion of the latter and their energy levels. The connection of these properties with the Periodic Table of the elements is examined.

Keywords Orbitals · Energy levels · Atomic number · Ionization potential · Ions

Summary A nucleus carrying a positive electric charge *Ze* attracts, mainly through the Coulomb potential, and binds around it *Z* electrons (all negatively charged) to form electrically neutral structures called *atoms.*[1] The simpler atom is that of hydrogen consisting of a proton and an electron trapped around it at a distance[2] $r = a_B$. The electron, being confined within a finite volume, exhibits discrete energy levels, as in Fig. (3.1d′), line 01′, given by [3.1, 3.8]

$$\varepsilon_{n-1} \approx -\frac{e^2}{2a_B}\frac{1}{n^2} \approx -13.6\frac{1}{n^2}\text{eV}, \quad n = 1, 2, 3, \ldots \tag{8.1}$$

(If instead of a proton of charge *e* we had a nucleus of charge *Ze*, then the radius would be $r = a_B/Z$ and the energy levels would be multiplied by Z^2). There are as many atoms as the existing stable (or long lived metastable) nuclei, i.e., about 270. However, as far as the electronic properties of atoms are concerned, the number of

---

[1] It is not uncommon to have $Z - \zeta$ electrons around the nucleus; then the structure carries overall a positive charge ζe and it is called a ζ^+ *cation*. It is also possible to have $Z + \xi$ electrons trapped around the nucleus, creating thus a structure of total negative charge equal to $-\xi e$ called a ξ^- *anion*. Cations and anions (collectively called *ions*) are metastable structures which tend to combine with each other or with electrons to form electrically neutral structures.

[2] Instead of a proton we may have a deuteron, i.e., a proton/neutron nucleus. As far as the electron is concerned, the only change this will bring, it will be a minute increase in the reduced mass of the system from 0.9995 $m_e \approx m_e$ to 0.9997 $m_e \approx m_e$

E. N. Economou, *A Short Journey from Quarks to the Universe*,
SpringerBriefs in Physics, 1, DOI: 10.1007/978-3-642-20089-2_8,

neutrons N in the nucleus is practically irrelevant (see footnote 2). Thus we count as different atoms those with different Z which are about 90 and which are arranged in a table called the *periodic table of the elements*. Atoms having a nucleus of the same Z but different N are termed *isotopes* of the same element. The most important property of an atom is the so-called first *ionization potential*, I_P, defined as the minimum energy required to extract an electron from it, when this atom is in its ground state. The ionization potential of the anion with $Z + 1$ electrons is called the *electron affinity*, E_A, of the Z atom. The quantities I_P and E_A give the energy scale of chemical reactions. The radius, r_a, of each atom in its ground state, although not well defined experimentally, is also of high interest.

In this chapter we shall obtain the order of magnitude of the quantities I_P, E_A, r_a by dimensional analysis. We shall also determine the angular dependence of the electronic stationary waves (see Appendix I), called *atomic orbitals*, associated with each of the discrete single electron level (such as those in (8.1)). The energy ordering of the atomic orbitals given in Table 8.1 allows the construction of the periodic table of the elements (see Table 3 at the inside of the backcover), which constitutes the foundation of Chemistry.

8.1 Size and Relevant Energy of Atoms

In the study of atoms the relevant universal physical constants, as it was mentioned before, are the following three,[3] e, $\hbar$, m_e; the number Z of protons in the nucleus, called *atomic number*, is also important. Hence, by dimensional considerations, we can conclude that the radius, r_a, of any atom is given by the formula,

$$r_a = c_a(Z)\, a_B, \tag{8.2}$$

since $a_B \equiv \hbar^2/m_e e^2$, the Bohr radius, is the only combination of e, $\hbar$, m_e with dimensions of length. As shown in Fig. 8.1, it turns out that the numerical factor $c_a(Z)$ is in a range between about 1 and 5. A qualitative explanation of the saw-like shape of the curve in Fig. 8.1 shall be provided later on when the stationary wave-functions (also called eigenfunctions or atomic orbitals) are examined. Again, by dimensional considerations, we obtain the following result for the first ionization potential:

$$I_P = c_p'(Z)\frac{e^2}{a_B} = c_P(Z)\frac{e^2}{r_a}; \quad e^2/a_B \equiv \hbar^2/m_e\, a_B^2 \equiv e^4 m_e/\hbar^2 = 27.2 \text{ eV}, \tag{8.3}$$

since $e^4 m_e/\hbar^2$ is the only combination of e, $\hbar$, m_e with dimensions of energy. In Fig. 8.2a I_P is plotted versus Z. The numerical factor $c_p'(Z)$ is obtained by dividing the values of I_P in eV by 27.2; it varies between 0.14 (for Cs) and 0.9 (for He). If

[3] The velocity of light c may also play a role, usually small; however, some properties, such as the magnetic properties, depend critically on c, and disappear in the electrostatic limit, $c \to \infty$.

Table 8.1 Electronic energy levels in non-hydrogenic atoms. The ordering of these levels and Pauli's principle determines the structure of the Periodic Table of the Elements (PTE)

ENERGY ↑				Number of electrons in the fully occupied shell	Total number of electrons up to this shell	
7s	7p	6d	5f	32	32+86=118	7th shell, 7th line
6s	6p	5d	4f	32	32+54=86	6th shell, 6th line
5s	5p	4d		18	18+36=54	5th shell, 5th line
4s	4p	3d		18	18+18=36	4th shell, 4th line
3s	3p			8	8+10=18	3rd shell, 3rd line
2s	2p			8	8+2=10	2nd shell, 2nd line
1s				2	2	1st shell, 1st line in the P.T.E
$s \Leftrightarrow \ell = 0$	$p \Leftrightarrow \ell = 1$	$d \Leftrightarrow \ell = 2$	$f \Leftrightarrow \ell = 3$			
2	6	10	14 ⇐	*Maximum number of electrons for each value of ℓ including spin*		

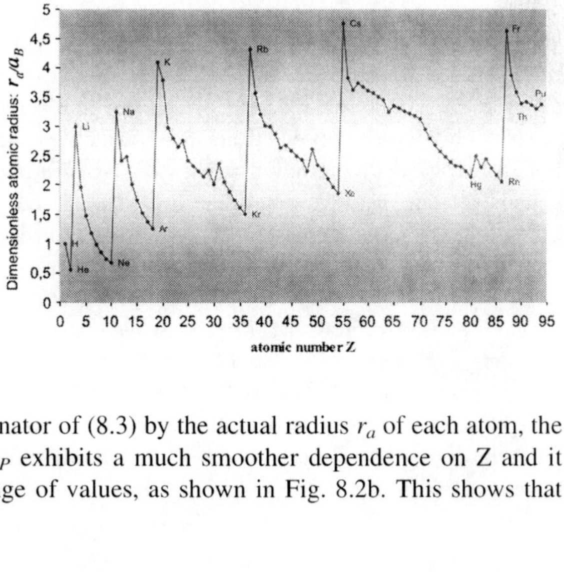

Fig. 8.1 Dimensionless radius of atoms, r_a/a_B versus their atomic number, Z. Local minima are associated with atoms of noble gases and local maxima with atoms of alkalis

we replace a_B in the denominator of (8.3) by the actual radius r_a of each atom, the resulting numerical factor c_P exhibits a much smoother dependence on Z and it varies in a more narrow range of values, as shown in Fig. 8.2b. This shows that

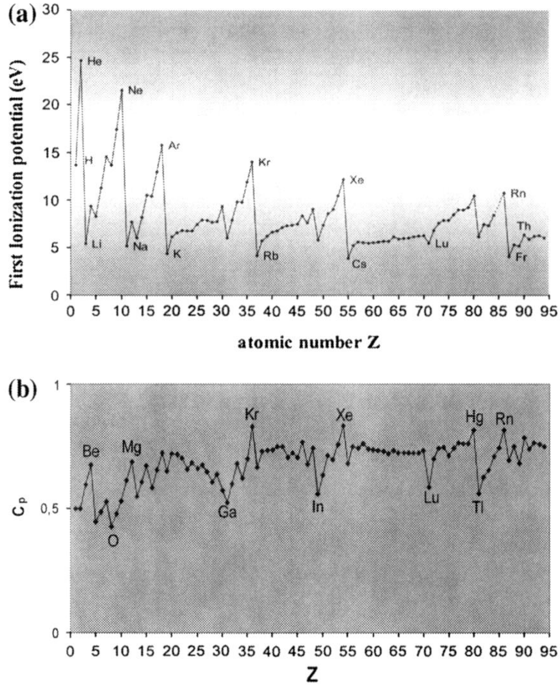

Fig. 8.2 **a** First Ionization Potential versus the atomic number Z. The local maxima are associated with completed shells, as in noble gases, and subshells. The local minima are associated mainly with alkalis and the Boron column of the periodic table. **b** The numerical factor $c_P(Z)$ versus Z (see (8.3)) shows a much smoother dependence on Z than I_P or $c'_p(Z)$ with an average value of $c_P = 0.68$

there is a strong correlation between I_P and r_a: the smaller the radius the larger the ionization potential and vice versa, as expected in view of the Coulomb potential which is inversely proportional to the length r.

8.2 Atomic Orbitals

To obtain the electronic properties of each atom with atomic number Z we have to find first the single electron stationary wavefunctions, the so-called atomic orbitals, and the corresponding discrete energy levels. Then we have to populate each atomic orbital with two electrons of opposite spin starting with the orbital of the lowest energy and continuing to the next ones in energy until all Z electrons are exhausted. This way we obtain the lowest total energy of the atom, the so-called ground state energy, consistent with Pauli's exclusion principle. Excited states of the atom can be obtained by redistributing the electrons among the atomic orbitals as to populate some that were empty in the ground state.

Each atomic orbital is fully determined by a wavefunction $\psi(\boldsymbol{r})$ which is a stationary solution (i.e., one of fixed energy) of a wave equation known as *Schrodinger's equation*; in the latter enters the Coulomb potential energy felt by an electron at position $\boldsymbol{r}$ due to the nucleus and *all the other electrons*. This potential energy, $V(r)$, as a result of the overall spherical symmetry of the atom, depends only on the distance r from the nucleus and not on the direction of $\boldsymbol{r}$. On the other hand, each individual stationary wavefunction, $\psi(\boldsymbol{r})$, may depend not only on the distance r but also on the angles θ, and φ which determine the direction of $\boldsymbol{r}$. Because $V(r)$ depends only on r, we can write $\psi(\boldsymbol{r})$ as a product, $\psi(\boldsymbol{r}) = R(r)Y(\theta, \varphi)$, with the angular part $Y(\theta, \varphi)$ being independent of the form and the values of $V(r)$. By choosing, $V(r) = E$, Schrodinger's equation reduces to the so-called Laplace equation,

$$\left(\partial^2\psi/\partial x^2\right) + \left(\partial^2\psi/\partial y^2\right) + \left(\partial^2\psi/\partial z^2\right) = 0, \quad (8.4)$$

the angular part of which is the same as the angular part of Schrodinger's equation. Hence, all we have to do is to find solutions of (8.4) and extract from them their angular part, which is the same as that of the atomic orbitals.

It is very easy to find polynomial solutions of degree ℓ for (8.4). One such solution is a polynomial of zero degree, $\ell = 0$, i.e., a constant whose angular dependence is also a constant. The solutions of Schrodinger's equation of $\ell = 0$ i.e., of no angular dependence are atomic orbitals denoted by the letter s.

Next, we consider polynomial solutions of (8.4) of degree $\ell = 1$, denoted by the letter p; there are exactly three *independent* such solutions: $\psi \sim x$ with angular dependence $p_x = x/r = \sin\theta\cos\varphi$; $\psi \sim y$ with angular dependence $p_y = y/r = \sin\theta\sin\varphi$; and $\psi \sim z$ with angular dependence $p_z = z/r = \cos\theta$.

Next, we consider polynomial solutions of (8.4) of degree $\ell = 2$, denoted by the letter d. There are exactly five *independent* such solutions: $\psi \sim xy, yz, zx, x^2 - y^2$ and $y^2 - z^2$. Continuing like this we find that there are exactly seven independent solutions corresponding to $\ell = 3$ denoted by the letter f. In general, there are $2\ell + 1$ independent polynomial solutions of (8.4) of degree ℓ with all their terms being of the same degree ℓ.

We conclude that the angular part of the atomic orbitals are characterized by two integer numbers: the degree ℓ of the polynomial solution of (8.4) and a number m identifying each one of the $2\ell + 1$ angular parts belonging to the same ℓ. To fully characterize an atomic orbital we need a third integer, denoted by n_r, $n_r = 0, 1, 2, \ldots$, and called the radial quantum number; n_r determines the radial part $R(r)$; Usually, instead of n_r, another integer is used, called *principal quantum number*, $n \equiv n_r + \ell + 1$, $n = \ell + 1$, $\ell + 2$, $\ell + 3$, ... In Fig. 8.3 we present a few atomic orbitals to indicate their angular dependence and to show how their relative size depends on the quantum numbers n and ℓ.

8.3 Energy Ordering of the Levels Corresponding to the Atomic Orbitals ψ_{nlm} and the Structure of the Periodic Table of the Elements

We have already stated that the energy levels of the hydrogen atom depend only on the principal quantum number n (see (8.1)). This is a peculiarity of the $1/r$ potential appropriate for the hydrogen atom. For all other atoms, which have more than one

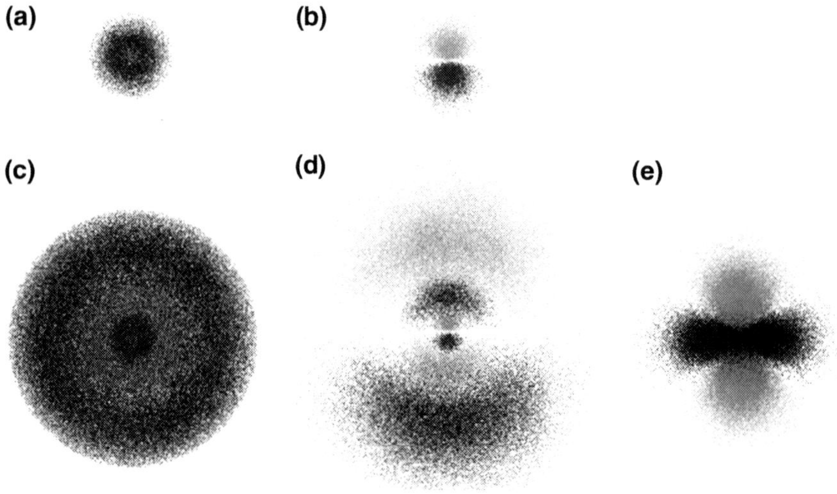

Fig. 8.3 Various atomic orbitals. Orange color (online, lighter grey in print) shows positive values of ψ and blue negative. In the following notation the number indicates the principal quantum number n and the letter indicates the angular dependence. **a** 2s; **b** $2p_z$; **c** 4s; **d** $4p_z$; *e* $3d_{z^2-\frac{1}{3}r^2}$. The **c** and the **e** orbitals, in spite of their difference in size, correspond to comparable energies. Note the increase in the size of the orbitals with n

electrons, the potential $V(r)$ is more complicated and, as a result, the energy level corresponding to the atomic orbital ψ_{nlm} depends on both n and ℓ, but not on m. This last non-dependence is a consequence of the fact that the rigid rotation of an atomic orbital around an axis passing through the nucleus does not change its energy, since there is no preferred direction in an isolated atom. If a preferred direction is introduced by placing the atom in an external field, then a dependence on m as well is expected to appear. The energy ordering of the atomic levels is very important, because it allows the identification of which orbitals are occupied in the ground state of the atom and which is the highest occupied orbital. The latter is of special importance in chemistry, because usually it is the most external one and, hence, the first one to come in contact with another atom (or atoms) to form molecules.

The rules of energy ordering of the atomic orbitals are relatively simple:

1. For the same principal quantum number n the higher ℓ has the higher energy: $\varepsilon_{n\ell} < \varepsilon_{n\ell'}$, if $\ell < \ell'$. This follows from the observation that higher ℓ (for the same n) is associated with the electron being further away from the nucleus (compare C and D in Fig. 8.3) and, hence, weaker attraction.
2. The orbitals $(n + 2, s)$ $(n + 1, d)$, and (n, f), have atomic levels of comparable energy. The energy ordering among these orbitals depends on their relative occupation by electrons. As a result, the ground state occupation of the atomic orbitals for the transition and the rare earth elements, which do not have fully occupied d and f orbitals, is not always a priori obvious.

3. The level of the orbital $(n + 2, p)$ is always higher than the levels of the orbitals $(n + 1, d)$ and (n, f) but lower than the orbital $(n + 3, s)$.

These rules allow us to energy-order the various atomic orbitals, with the uncertainties mentioned in point (2) above, as shown in Table 8.1 and to construct line by line and shell by shell the periodic table of the elements (PTE).

Let us present, as an example, the construction of the fourth line of the PTE, from Z = 19 (K) till Z = 36 (Kr): and ignoring the already fully occupied 1*s*, 2*s*, 2*p*, 3*s*, and 3*p* orbitals. The first two electrons go to the orbital 4*s* completing thus the *s*-subshell at the element Ca (Z = 20). The next three electrons go to the 3*d* orbital indicating that up to this occupation the 4*s* orbital is lower in energy than the 3*d*; we have this way the elements: Z = 21, Sc, with $4s^2 3d^1$; Z = 22, Ti, with $4s^2 3d^2$ and Z = 23, V with $4s^2 3d^3$. The exponents indicate the number of electrons in the corresponding orbital. In the next element Z = 24, Cr, the first reversal of the ordering appears, since the configuration is not $4s^2 3d^4$, but $4s^1 3d^5$ indicating that at this stage of occupation the 3*d* orbital is lower in energy than the 4*s*. In the next four elements, Z = 25, Mn, Z = 26, Fe, Z = 27, Co, and Z = 28, Ni the original ordering returns: $4s^2 3d^5$, $4s^2 3d^6$, $4s^2 3d^7$, and $4s^2 3d^8$ respectively. In the next element Z = 29, Cu, the anomaly returns: the configuration is $4s^1 3d^{10}$. With the next element, Z = 30, Zn, the *s/d* subshell is completed. The next six elements, from Z = 31, Ga to Z = 36, Kr, correspond to the gradual occupation of the 4*p* orbitals.

Now, let us go back to Figs. 8.1 and 8.2 and examine how the atomic radius and the first ionization potential vary as we go along the elements of the fourth line of the PTE. The first observation is the rather systematic drop of the atomic radius and the corresponding increase of the ionization potential. This trend is due to the fact that the first electron of a new shell is loosely bound, while all the electrons of a completed shell are strongly bound, since a large energy difference separates the orbital *n*, *p* from the *n* + 1, *s*. Notice the non-monotonic behavior appearing at Z = 30 and Z = 31. The element Z = 30 behaves somehow similar to that of Z = 36 in the sense that it exhibits a local minimum of the atomic radius and a local maximum of the I_P; this is because the *s*/*d* subshell is completed for Z = 30. In contrast, the element Z = 31, exhibits a local maximum of the atomic radius and a local minimum of the I_P. In this sense it is similar to Z = 19, K, because it has only one electron in a new subshell. The non-monotonic behavior of the atomic radius appearing between Z = 23 and Z = 24 and between Z = 28 and Z = 29 is due to the anomaly of the orderings between the 4*s* and 3*d* orbitals.

Chapter 9
From Atoms to Molecules

Abstract Atoms attract each other to form molecules. For each molecule we would like to know the relative positions of its atoms, their vibration properties, and the changes in the electronic motion induced by molecular formation. Dimensional analysis and various computational methods such as the so-called LCAO are useful tools in obtaining this information.

Keywords Bond length · Dissociation energy · Vibrations · Rotations · LCAO · Hybridization

Summary The atoms come close together to form molecules, because by doing so they lower their energy, as shown in Fig. 9.1. There is an unlimited number of different molecules starting from simple diatomic molecules, such as N_2 or Nacl, and reaching the huge biomolecules consisting of millions of atoms. The so-called *bond length*, that is the equilibrium distance d between the nuclei of neighboring atoms in a molecule, is of the order of the sum of the atomic radii, $d \approx r_{a1} + r_{a2}$, since neighboring atoms touch each other. For each molecule we would like to find a number of properties such as its stereochemistry (i. e., the relative position of every one of its atoms in space), the spectrum of the discrete frequencies and the corresponding stationary states of ionic vibrations around their equilibrium positions,[1] as well as the electronic stationary states (i.e., the molecular orbitals, occupied and unoccupied, when the molecule is in its ground electronic state). In this chapter we shall first explain, by means of the important graph 9.1 (Fig. 9.1), why atoms form molecules. Then we shall employ dimensional considerations to estimate the values of various quantities pertaining to molecules. A method known as LCAO, expressing each molecular orbital as a linear

---

[1] Each atom in a molecule is conceptionally divided into the external valence electrons responsible for the bond and the remaining ion, the electrons of which are unaffected by the molecule formation.

E. N. Economou, *A Short Journey from Quarks to the Universe*, SpringerBriefs in Physics, 1, DOI: 10.1007/978-3-642-20089-2_9,

combination of atomic orbitals, will be introduced. It will be used in order to make an educated guess regarding the form of the molecular orbital, and to estimate the corresponding molecular electronic levels, and the stereochemistry of, at least, some simple molecules. Finally, the idea of hybridization of atomic orbitals will be presented.

9.1 The Residual Electric Interaction Between Two Atoms

Consider a system of two atoms at a distance d' between their nuclei. It is important to obtain the interaction energy, E, of this system as a function of d', when the latter varies between, let us say, 20 and 0.001 Å. This energy includes the Coulomb contributions among all the charged particles of both atoms and the kinetic energy of all the electrons; the latter are assumed to be in their lowest total energy for every distance d'. The kinetic energy of the two nuclei is omitted for the time being. The dependence of E on d' is shown schematically in Fig. 9.1 and its gross characteristics can be justified as follows: At a distance d' much larger than the equilibrium distance d, one can show that E is inversely proportional to the sixth power[2] of d':

$$E = -A/d'^6, \text{ with } A > 0,\ d' \gg d; \quad \text{van der Waals interaction.} \tag{9.1}$$

On the other hand, at d' much smaller than d, the Coulomb repulsion between the two nuclei will dominate[3] giving a contribution to E as shown in Fig. 9.1 with q'^2 approaching $Z_1 Z_2 e^2$. Since the E versus d' curve starts being negative and decreasing with decreasing d' and ends up being positive and increasing, it must have at least one negative minimum for some intermediate value of d'. It turns out that there is only one minimum occurring when the highest occupied orbitals of each atom have a small overlap. If the two atoms are squeezed together beyond this point, a repulsion develops partly due to the increased overlap of the occupied atomic orbitals and partly due to the Coulomb repulsion of the two nuclei. Actually

[2] The interaction energy E of of two neutral system at a distance d' much larger than their size is proportional to $-\boldsymbol{E}_1(\boldsymbol{d}')\cdot\boldsymbol{p}_2$, where $\boldsymbol{E}_1(\boldsymbol{d}') \sim \boldsymbol{p}_1/d'^3$ is the electric field created by the dipole moment $\boldsymbol{p}_1$ at a distance d'; $p_2 = \alpha_2 \cdot E_1(d')$ is the dipole moment of the system 2 induced by the field $\boldsymbol{E}_1(\boldsymbol{d})$ and α_2 is the polarizability of the system 2. For dimensional reasons, α_2 is proportional to the volume r_{a2}^3 (see Sect. 5.3); the latter can be written with the help of (8.3) and $p_2^2 \sim e^2 \cdot r_{a2}^2$ as $r_{a2}^3 = r_{a2}^2 r_{a2} = r_{a2}^2 c_{p2} e^2/I_{P2} \propto p_2^2/I_{P2}$. Substituting $\boldsymbol{E}_1(d')$, $\boldsymbol{p}_2$, and α_2 in $E = -\boldsymbol{E}_1(d')$, $\boldsymbol{p}_2$ we obtain $E = -A/d'^6$, where $A \sim p_1^2 p_2^2/I_{P2}$. If we have started with the equivalent relation $-E_2(d')\cdot p_1$ for E instead of $-E_1(d')\cdot p_2$, the result for A would have been $A \sim p_1^2\, p_2^2/I_{p1}$; it is not unreasonable to assume, for symmetry reasons, that the correct expression for A is the average of the two: $A = c_W p_1^2\, p_2^2\left(\frac{1}{I_{P1}} + \frac{1}{I_{P2}}\right)$. For the hydrogen–hydrogen case, where $p_1^2 = p_2^2 = 3e^2 a_B^2$, the numerical factor c_W is equal to 0.18.

[3] The electrons will not be squeezed between the two nuclei as to screen their repulsion; on the contrary, they will approach the ground state configuration of an atom of atomic number $Z_1 + Z_2$, where Z_1, Z_2 are the atomic numbers of the two atoms under consideration.

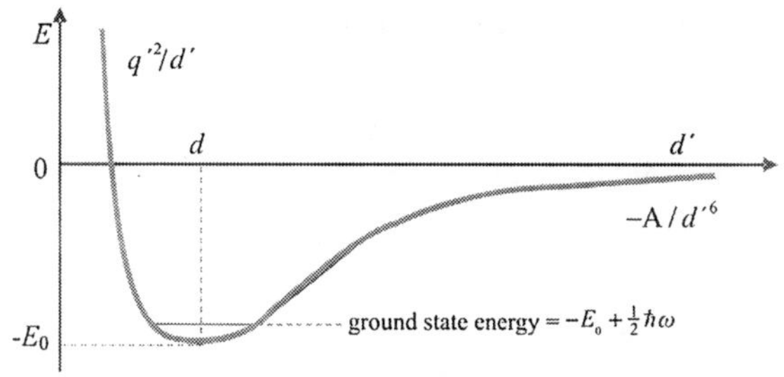

Fig. 9.1 Schematic plot of the variation of the total energy E of two neutral atoms, versus the distance d' between the two nuclei. The ground state energy of the two atoms when $d' = \infty$ is chosen as the zero of energy. The zero point motion of the ions increases the minimum energy at the equilibrium distance by $\frac{1}{2}\hbar\omega$.

what happens as the two atoms start approaching each other from large distance, there is first a slight reorganization of the external electronic states within each atom giving rise to the van der Waals attraction. When the atoms start touching each other the reorganization is more drastic and involves the spreading of the external electrons to the other atom, i.e., the change from atomic orbitals to molecular orbitals[4] takes place.

The E versus d' curve allows us to extract a number of valuable information regarding diatomic molecules: For example, the equilibrium distance d gives the bond length; actually there is a fluctuation Δd of this length, since otherwise Heisenberg's relation $\Delta d \cdot \Delta p_d \geq \hbar/2$ would be violated. This fluctuation gives rise to an ionic kinetic energy $\hbar^2/8m_r\Delta d^2((m_r \equiv m_{a1}m_{a2})/(m_{a1}+m_{a2}))$, and an additional potential energy equal to $\frac{1}{2}\kappa\,\Delta d^2$ where $\kappa = (\partial^2 E/\partial d'^2)_{d'=d}$. Minimizing the sum ε_{vo} of these two energies (due to the minimum relative motion of the two ions and known as *zero point motion*) we obtain that $\Delta d^2 = \hbar/2m_r\omega$, where $\omega \equiv \sqrt{\kappa/m_r}$. Substituting this value of Δd in the sum of the kinetic and the potential energies we find that $\varepsilon_{vo} = \frac{1}{2}\hbar\omega$.

To summarize: From the E versus d' dependence we can extract the following information: (1) The bond length d and, hence, the rotational spectrum $\varepsilon_{r\ell} = \hbar^2\ell(\ell+1)/2I$, where $\hbar^2\ell(\ell+1)$ is the square of the angular momentum, $\ell = 0, 1, 2, \ldots$, and $I = m_r d^2$ is the moment of inertia of the molecule. (2) The fluctuation in the bond length Δd. (3) The natural frequency of oscillation ω and, hence, the vibrational spectrum $\varepsilon_{vn} = (n+\frac{1}{2})\hbar\omega$, $n = 0, 1, 2, \ldots$ (4). The dissociation energy (i.e., the minimum energy needed to separate the molecule to the two atoms), $D = |E_o| - \frac{1}{2}\hbar\omega$.

[4] Atoms of the noble gases are an exception: Because of their fully completed shells and the large energy separation of the next empty level, no overlap of atomic orbitals belonging to different atoms is tolerated, and the curve E versus d' starts moving upwards before any overlap occurs and before any molecular orbital is formed. As a result, the equilibrium distance d is considerably larger than the sum $r_{a1} + r_{a2}$ and the energy gain $|E_o|$ is much smaller than usually.

9.2 Estimates Based on Dimensional Analysis

Some of the molecular quantities mentioned in the previous section can be estimated by dimensional analysis as follows:

1. Bond length d (see Fig. 9.1 and footnote 3 in this chapter)

$$d \approx (r_{a1} + r_{a2}) = (c_{a1} + c_{a2})a_B \approx 0.5(c_{a1} + c_{a2})\text{A} \tag{9.2}$$

The argument in support of this relation was presented in the previous section.

2. *Dissociation energy D*

$$D = c_d\hbar^2/m_e d^2 = c_d(\hbar^2/m_e a_B^2)(a_B^2/d^2) \approx c_d 27.2(a_B^2/d^2)\text{eV} \tag{9.3}$$

It is expected that the order of magnitude of the dissociation energy must be given by the basic unit of energy, $\hbar^2/m_e a_B^2$. Moreover, it is reasonable to expect that the replacement of a_B by the characteristic length d of the molecule will make our estimate molecule-specific, the same way that the replacement $a_B \to r_a$ made our estimate of I_P atom-specific (see Fig. 7.2). The quantity D for the molecules O_2, Rb_2, AgI is 5.17, 0.51, 2.42 eV, respectively. To obtain these values from (9.3) and the corresponding experimental data for d, $d = 1.207$, 4.4, 2.54 Å, we must choose c_d equal to 1.028, 1.29, 2.05 respectively. Notice that (9.2) and Fig. 8.1 give $d = 0.9$, 4.49, 2.64 Å, respectively.

3. *Natural vibrational frequency* ω. As explained in Chap. 5, the vibrational frequency ω must be proportional to ω_o, where $\omega_o = e^2/a_B\hbar$ is the only quantity with dimensions of frequency made out of e, m_e, $\hbar$. However, the vibrating masses are those of the ions; thus the reduced mass of the ions must enter the formula for ω and it must enter in a dimensionless way. Hence the most general formula for ω is of the following form: $\omega = \omega_o \text{f}(m_r/m_e)$, where f is an arbitrary function of m_r/m_e. We demonstrate in App. I that in an oscillation, the oscillating mass enters as the inverse square root. Therefore, by implementing also the replacement $a_B \to d$, the formula for ω becomes

$$\omega = c_v \frac{e^2}{d\hbar}\sqrt{\frac{m_e}{m_r}} \tag{9.4}$$

where the numerical factor is in the range between 0.5 and 2.5. Choosing $c_v = 1$, and the experimental values for d, (9.4) gives that $\hbar\omega$ equals 453, 116, 98.7, and 32,3 meV for H_2, N_2, O_2, and Na_2, respectively, while the corresponding experimental values are 546, 292, 196, and 19.7, respectively.

4. *The rotational spectrum.* As it was mentioned in the previous section, the quantum of the rotational spectrum of a diatomic molecule is equal to

$$\hbar^2/I = \hbar^2/m_r d^2 = (\hbar^2/m_e a_B^2)(m_e/m_r)(a_B^2/d^2) = 27.2(m_e/m_r)(a_B^2/d^2)\text{eV} \tag{9.5}$$

For O_2 the quantity $\hbar^2/I$ is equal to 0.36 meV. In general the rotational quantum (such a quantum is what we called $\delta\varepsilon$ in Chap. 3) is smaller than the typical dissociation energy D, by a factor (m_e/m_r). Its range of values is about a fraction of a meV which corresponds to a few degrees Kelvin. On the other hand, the vibrational quantum $\hbar\omega$ is smaller than the typical electronic energy by a factor $\sqrt{m_e/m_r}$ which brings it down from a few eV to the few tens (even hundreds) of meV s (corresponding to hundreds or even thousands of degrees Kelvin).We conclude that at room temperature the rotational degrees of freedom of a diatomic molecule are fully excited, while the vibrational ones are usually frozen, at least for small, light molecules.

Out of the $3N_a$ ionic degrees of freedom of a non-linear molecule consisting of N_a atoms, three degrees of freedom are associated with the motion of the center of mass of the molecule and another 3 are associated with its rotations as a rigid body.[5] The remaining $3N_a - 6$ ionic degrees of freedom are associated with stationary oscillations (or waves, see Fig. Ic in p. 126) to each one of which corresponds a natural frequency common to all participating ions.

9.3 Linear Combination of Atomic Orbitals

The LCAO method attempts to approximately calculate the molecular orbitals by expressing them as linear combination of atomic orbitals (see the first few tens of pages of reference [9.1]). To demonstrate the LCAO method we shall consider first the simple molecule AgI and we shall assume that only two atomic orbitals, the $5s$ of Ag (denoted by ψ_1), and the $5p_x$ of I (denoted by ψ_2), are relevant for the formation of the molecule. This assumption, although oversimplified, is not unreasonable, since these two atomic orbitals are the highest singly occupied ones (each of them has just one electron). Then, according to the LCAO method, and the above simplifying assumption, the molecular orbital ϕ is given by

$$\phi = c_1\psi_1 + c_2\psi_2 \tag{9.6}$$

The physical meaning of the coefficients c_1, c_2 is the following: If an electron is in the molecular orbital ϕ the probability p_1 to find it in the atomic orbital ψ_1 is given by $p_1 = c_1^2/(c_1^2 + c_2^2)$, and the probability to find it in the atomic orbital ψ_2 is given by $p_2 = c_2^2/(c_1^2 + c_2^2)$. The average value of the energy of an electron being in the orbital ϕ can be obtained from the following general formula[6]:

---

[5] If the molecule is linear, there are only two rotational degrees of freedom and $3N-5$ vibrational ones.

[6] If a waveparticle is in a state ϕ, the average value of a physical quantity A is obtained by the following formula: $\langle\langle A\rangle = \langle\phi|A|\phi\rangle/\langle\phi|\phi\rangle\rangle$ where by definition we have $\langle\phi|A|\phi\rangle \equiv \int \phi A\phi d^3r$ and $\langle\phi|\phi\rangle \equiv \int \phi\phi d^3r$; ϕ, c_1, c_2 are assumed to be real.

$$\varepsilon_\phi = \langle\phi|H|\phi\rangle/\langle\phi|\phi\rangle = (c_1^2\varepsilon_1 + c_2^2\varepsilon_2 + 2c_1c_2V_2)/(c_1^2 + c_2^2) \tag{9.7}$$

where $\varepsilon_1 \equiv \langle\psi_1|H|\psi_1\rangle$, $\varepsilon_2 \equiv \langle\psi_2|H|\psi_2\rangle$, $V_2 \equiv \langle\psi_1|H|\psi_2\rangle$ and $H = p^2/2m_e + V(\mathbf{r})$. We can determine c_1 and c_2 by minimizing ε_ϕ. More specifically, we determine the extrema of ε_ϕ by setting the derivatives $\partial\varepsilon_\phi/\partial c_1, \partial\varepsilon_\phi/\partial c_2$ equal to zero. We thus find the following homogeneous linear system:

$$(\varepsilon_1 - \varepsilon_\phi)c_1 + V_2c_2 = 0, \quad (\varepsilon_2 - \varepsilon_\phi)c_2 + V_2c_1 = 0 \tag{9.8}$$

which has two solutions. The one with the lower energy is

$$\varepsilon_\phi = \varepsilon_b = \varepsilon - \sqrt{V_2^2 + V_3^2}, \text{ where } \varepsilon \equiv (\varepsilon_1 + \varepsilon_2)/2, \text{ and } V_3 \equiv (\varepsilon_1 - \varepsilon_2)/2 > 0 \tag{9.9}$$

with the values of c_1, c_2 of the corresponding orbital ϕ_b such that

$$p_{1b} = c_1^2/(c_1^2 + c_2^2) = (1 - a_p)/2 \text{ and } p_{2b} = c_2^2/(c_1^2 + c_2^2) = (1 + a_p)/2 \tag{9.10}$$

where the so-called *polarity index* a_p is defined as follows:

$$a_p \equiv V_3/\sqrt{V_2^2 + V_3^2} > 0 \tag{9.11}$$

If we place the two valence electrons in the molecular orbital ϕ_b with opposite spins, their total energy will be $2\varepsilon_b = 2\varepsilon - 2\sqrt{V_2^2 + V_3^2}$, while before the molecule formation the corresponding energy was $\varepsilon_1 + \varepsilon_2 \equiv 2\varepsilon$. Hence, there is a lowering of energy due to the molecule formation equal to $2\sqrt{V_2^2 + V_3^2}$. Actually the lowering is $2\sqrt{V_2^2 + V_3^2} - U$, because the presence of the two electrons in the same orbital leads to an increase U of energy relative to the situation before where the two electrons were in different atoms. We conclude that placing the two electrons to the orbital ϕ_b lowers the total energy by $D = 2\sqrt{V_2^2 + V_3^2} - U - \frac{1}{2}\hbar\omega$, and, hence, bonds the two atoms together as to form a diatomic molecule. For this reason, ϕ_b is called *bonding molecular orbital* and the level ε_b is called bonding molecular level. Equation (9.10) shows that out of the two electrons populating the orbital ϕ_b $2p_{2b} = 1 + a_p$ electrons are on the average in the atom 2 and $2p_{1b} = 1 - a_p$ electrons are in the atom 1. Therefore, the molecule formation is accompanied by a net transfer of electron equal to a_p from atom 1 (with the higher ε_1) to atom 2 (with the lower ε_2); this justifies the name polarity index for a_p.

Let us now examine the other solution of the system (9.8). The resulting molecular orbital, denoted by ϕ_a, is given by

$$\phi_a = \frac{1}{\sqrt{2}}\left(\sqrt{1 + a_p}\psi_1 - \sqrt{1 - a_p}\psi_2\right) \tag{9.12}$$

and the corresponding molecular level is given by

$$\varepsilon_\phi = \varepsilon_a = \varepsilon + \sqrt{V_2^2 + V_3^2} \tag{9.13}$$

Placing two electrons in the orbital ϕ_a will increase now the energy by an amount equal to $2\sqrt{V_2^2 + V_3^2} + U'$. Hence, the molecular orbital ϕ_a works in the opposite direction than that of molecule formation and for this reason it is called the *antibonding molecular orbital*; ε_a is called the antibonding molecular level.

What would have happened, if the atomic levels ψ_1, ψ_2 were fully occupied with two electrons each? In that case, we would be forced to place two electrons in the bonding molecular orbital, ϕ_b, and two in the antibonding ϕ_a; the result would be to increase the energy by an amount about equal to $U + U'$, instead of lowering it. We conclude that fully occupied atomic orbitals do not favor molecule formation through molecular orbitals and, as a first approximation, can be ignored (see footnote 4 p. 71). The ones that favor molecule formation are those which are singly occupied.

Up to now the presentation of the LCAO method was kept at a qualitative level. To reach a semiquantitative level we need to know the atomic levels ε_1 and ε_2 and the value of V_2, as well as an estimate of U. The levels ε_1 and ε_2 are given approximately in Table 4 in p. 144, while for V_2 the approximate expression proposed by Harrison (9.2) will be employed. According to this:

$$V_2 = c_2 \hbar^2 / m_e d^2 \tag{9.14}$$

with $c_2 = -1.32,\ 1.42,\ 2.22,\ -0.63$ for s/s, s/p_x, p_x/p_x, p_y/p_y atomic orbitals respectively, where the x-axis is taken along the axis of the molecule. U is usually taken as $c_U\sqrt{V_2^2 + V_3^2}$ with c_U in the range between about 1 and 1, 5; the smaller values for single long bonds and the larger for triple short bonds. For the molecule AgI we have the following values $\varepsilon_1 = -5.98$ eV, $\varepsilon_2 = -10.97$ eV, $d = 4.8 \times 0.53 = 2.54$ Å, $V_2 = 1.68$ eV, $V_3 = 2.495$ eV, $D \approx 3.00$ eV by choosing $c_U = 1$, versus $D = 2.42 \pm 0.3$ eV for the experimental value.

For diatomic molecules consisting of identical atoms, the above formulae become simpler, because then $V_3 = a_p = 0$ and $c_1^2/(c_1^2 + c_2^2) = c_2^2/(c_1^2 + c_2^2) = 1/2$. As an example the molecule Na_2, for which the relevant atomic orbital is the $3s$ for each atom, has $d = 3.08$ Å. Then we find: $V_2 = -132\,\hbar^2/m_e d^2 = 1.06\,\text{eV}$ and, by choosing $c_U \approx 1, D = |V_2| \approx 1$ eV, while the experimental value is $D = 0.78$ eV. For the molecule O_2 there are two *independent* pairs of relevant atomic orbitals giving rise to a double bond. The p_{x1}/p_{x2} (with $V_{2x} = 2.22\,\hbar^2/m_e d^2 = 11.58\,\text{eV}$) produces a strong bond and the p_{y1}/p_{y2} (with $V_{2y} = -0.63\,\hbar^2/m_e d^2 = 3.29\,\text{eV}$), a weak one. The dissociation energy in this case is given by the following formula: $D = (2 - c_U)(|V_{2x}| + |V_{2y}|) = 7.28$ eV, where in the numerical results we have taken the experimental value of $d = 1.208$ A and we have chosen $c_U \approx 1.5$. For the nitrogen molecule there are three independent pairs making the triple bond: p_{x1}/p_{x2} (with $V_{2x} = 2.22\,\hbar^2/m_e d^2 = 14\,\text{eV}$), p_{y1}/p_{y2} (with $V_{2y} = -0.63\,\hbar^2/m_e d^2 = -4\,\text{eV}$), and p_{z1}/p_{z2} (with $V_{2z} = V_{2y}$). By choosing $c_U = 1.5$ and using the actual value of $d = 1.098$ Å, we find $D = 11$ eV compared to $D = 9.79$ eV measured experimentally.

9.4 Hybridization of Atomic Orbitals

In the applications of the LCAO method we have presented up to now only the singly occupied atomic orbitals have been evolved. Moreover, for the O_2, N_2 molecules the four or six relevant atomic orbitals were naturally grouped into two or three *independent* pairs. Thus the calculation was reduced to simply solving systems of two linear homogeneous algebraic equations. In general, nature is more complex. To see why consider the important molecule CO_2. If we employ only singly occupied atomic orbitals, this molecule will look as follows:

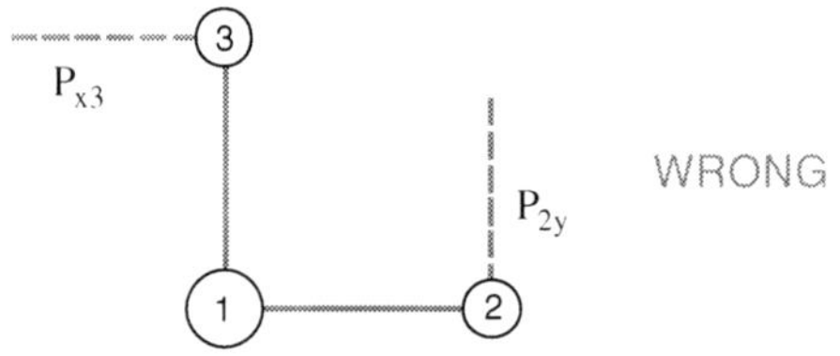

where the bonding molecular orbital (12) between the carbon atom (no. 1) and the oxygen atom (no. 2) will involve the pair of atomic orbitals p_{x1} and p_{x2}, while the (13) bond will involve the pair p_{y1}, p_{y3}. The problem is that the oxygen orbitals p_{y2} *and* p_{x3} are left dangling and, hence, their possibility to form bonds was not exploited. Such a configuration of CO_2 would be unstable. The only way out of this problem is to use the $2s$ atomic orbital of carbon which in combination with its three $2p$ orbitals will make four atomic orbitals with four electrons to pair with the four orbitals of the two oxygens (p_{x2}, p_{y2}, p_{x3}, p_{z3}) with their four electrons. The calculation now is considerably more complicated than before: We have to find, by minimizing the energy, which linear combinations of the four atomic orbitals (s, p_x, p_y, p_z) of carbon will produce the ground and the excited states of the CO_2. To avoid the calculation of the total energy, setting its derivatives equal to zero, and solving the resulting system of eight equations with eight unknowns (four for the carbon orbitals and four for the orbitals of the two oxygens), a drastic approximation has been devised. This approximation consists in having three standard preselected combinations, called *hybridizations*, of the four atomic orbitals (one s and three p's) and for each case to check which one of the three seems to fit best. The three standard preselected hybridizations are the following (see the first few tens of pages of [9.1]):

1. $sp^{(1)}$: *There are two hybrid orbitals*: $\chi_1 = \frac{1}{\sqrt{2}}(s + p_x)$, $\chi_2 = \frac{1}{\sqrt{2}}(s - p_x)$ and two unhybridized ones, p_y, p_z; χ_1 is oriented along the positive x direction, while χ_2 along the negative x direction. The energy associated with either χ_1 or χ_2 is $\varepsilon_h = (\varepsilon_s + \varepsilon_p)/2$. There are two drawbacks associated with the hybridization: The first one is that there is a non-zero matrix element between χ_1 and χ_2 equal to $V_1 = -(\varepsilon_p - \varepsilon_s)/2$. The second is that there is an energy cost equal to $\varepsilon_s + \varepsilon_p + 2\varepsilon_p - 2\varepsilon_s - 2\varepsilon_p = \varepsilon_p - \varepsilon_s$. The latter must be viewed as an energy investment to be more than recovered with the molecule formation. The $sp^{(1)}$

hybridization is the appropriate one for the CO_2. One C–O bonding molecular orbital is made from the pair χ_1, p_{x2} with $V_2 = 2.57\,\hbar^2/m_e d^2$ and another from the pair p_{y1}, p_{y2} with $V_2 = -0.63\,\hbar^2/m_e d^2$ both to the right of the carbon atom. Two similar bonds, one strong and one weak, are formed to the left of carbon between the atomic orbitals χ_2, p_{x3} and p_{z1}, p_{z3}. Thus CO_2 is a linear molecule: O=C=O.

2. $sp^{(2)}$: There are three hybrid atomic orbitals, $\chi_1 = \frac{1}{\sqrt{3}}(s + \sqrt{2}p_x)$, and χ_2, $\chi_3 = \frac{1}{\sqrt{3}}\left(s - \frac{1}{\sqrt{2}}p_x \pm \frac{\sqrt{3}}{\sqrt{2}}p_y\right)$ each with energy $\varepsilon_h = (\varepsilon_s + 2\varepsilon_p)/3$. These three hybrids are in the same plane and the angle between two successive ones is 120°. The off-diagonal matrix element between any pair of the three hybrids is $V_1 = -(\varepsilon_p - \varepsilon_s)/3$. The fourth orbital, p_z, is unhybridized. The energy cost of the $sp^{(2)}$ hybridization is again $\varepsilon_p - \varepsilon_s$. An important example of $sp^{(2)}$ hybridization is the benzene molecule, C_6H_6, (see Pr. 8.5, and its solution, where the C_6H_6's stereochemistry is presented). The benzene is a planar ring molecule forming a regular hexagon; the six unhybridized p_z atomic orbitals of carbon in C_6H_6 combine to give rise to six *delocalized* molecular orbitals of the form:

$$\phi_v = \sum_{n=1}^{n=6} c_n^{(v)}\psi_n,\ c_n^{(v)} = c_o \exp(i\varphi_v n) \text{ and } \varphi_v = (2\pi/6)v,\ v = 0, \pm 1, \pm 2, 3. \tag{9.15}$$

The energy corresponding to each molecular orbital ϕ_v is $\varepsilon_v = \varepsilon_p + 2V_{2,zz}\cos\varphi_v$. The formation of these delocalized molecular orbitals gives an additional energy reduction besides that associated with the $sp^{(2)}$ hybrids, which is equal to $(4|V_{2,zz}|/3)$ per p_z orbital. Notice that the formation of three independent bonds, involving two p_z orbitals each, would give a smaller energy reduction equal to $|V_{2,zz}|$ per p_z orbital (the U term has been omitted). Thus delocalization pays off.

3. $sp^{(3)}$: All four atomic orbitals are hybridized, $\chi_1 = \frac{1}{2}(s + p_x + p_y + p_z)$, $\chi_2 = \frac{1}{2}(s - p_x - p_y + p_z)$, χ_3 and χ_4 are obtained from χ_2 by cyclic permutations of the two minuses. The directions of these hybrids are from the center of a regular tetrahedron to its four vertices. It follows that the cosine of the angle between any two of these four directions is $-1/3$ and the angle is 109.47°. The energy of each of these four hybrids is $\varepsilon_h = (\varepsilon_s + 3\varepsilon_p)/4$ and the off-diagonal matrix element is $V_1 = -(\varepsilon_p - \varepsilon_s)/4$. The energy cost of this hybridization is again $\varepsilon_p - \varepsilon_s$. Important molecules, such as methane, CH_4, ethane, C_2H_6, silane, SiH_4, etc., employ the $sp^{(3)}$ hybridization of C or Si (See Problem 8.3).

Chapter 10
From Atoms (or Molecules) to Solids (or Liquids)

Abstract Solids are made from a huge number of atoms close-packed, quite often periodically arranged, and undergoing small vibrations around their equilibrium positions. The same information as in molecules are sought in solids in order to deduce their various macroscopic properties. Similar theoretical and calculational tools as in molecules are employed to obtain quantities such density, compressibility, velocity of sound, specific heat, conductivity, etc.

Keywords Condensed matter · Periodicity · Bloch theorem · Defects · Bulk and Shear moduli

Summary Atoms (or molecules) come close together in huge numbers (A human body of M kg contains $N_a = 0.96 \times 10^{26} \times M$ atoms) to form solids, liquids, and states intermediate between solids and liquids, such as polymers; all these formations are collectively called *condensed matter*. In solids, atoms have fixed positions in space and undergo small oscillations around them. In contrast, each atom in liquids moves over the whole extent of the liquid, while in touch with whichever atoms happen to be around it. In crystalline solids a microscopic fixed cluster of atoms (or even a single atom) is repeated periodically in space. This periodic order, although not perfect, due to the unavoidable presence of defects, greatly facilitates the study of crystalline solids because it reduces it to the study of the cluster of atoms, thanks to the so-called *Bloch theorem*. (It is this theorem which led to the form shown in (9.15) for the coefficients $c_n^{(\nu)}$). Among the most important properties of condensed matter is its density, which is determined by the average atomic weight and the volume per atom, and its compressibility, the inverse of which is called *bulk modulus*, is denoted by B, and has dimensions of pressure. Solids, in contrast to liquids, put up resistance to stresses which tend to change their shape but not their volume. This property is quantified by the so-called *shear modulus* denoted by μ_s and having dimensions of pressure. Other mechanical properties of solids, such as failure by breaking or plastic deformation, are of obvious interest. Moreover, electrical properties such

E. N. Economou, *A Short Journey from Quarks to the Universe*,
SpringerBriefs in Physics, 1, DOI: 10.1007/978-3-642-20089-2_10,

as resistivity, magnetic properties such as ferromagnetism, optical properties such as index of refraction, thermal properties such as thermal conductivity, etc., are of great interest both from a point of view of basic science and from that of technology. In this chapter we extract a few basic properties of solids first by dimensional analysis and then by briefly presenting the so-called jellium model [10.1] and the LCAO method [9.2].

10.1 Dimensional Analysis Applied to Solids

As it was argued in the introduction of Part III (p. 59) the properties of condensed matter depend necessarily on e, $\hbar$, m_e, or, equivalently, on a_B, $\hbar$, m_e. They depend also on the type of atom(s) participating in its formation, that is on the atomic number Z (or Z's). They may depend also explicitly on the masses of the atoms, m_{ai}, on the temperature, T, on the pressure, P, on the velocity of light, c, etc. Following the general recipe for dimensional analysis presented in the summary of Chap. 5, we obtain the following general formula for any quantity X pertaining to the condensed matter:

$$\frac{X}{X_o} \equiv \bar{X} = f_x\left(Z_i, \frac{m_{\alpha i}}{m_e}(Z_i), \frac{T}{T_o}, \frac{P}{P_o}, \frac{c}{\upsilon_o}, \ldots\right), \tag{10.1}$$

where $\bar{X}$ is the value[1] of X in atomic units, X_o is the combination $X_o = \hbar^{\nu_1} m_e^{\nu_2} a_B^{\nu_3}$ which has the same dimensions as X (see Table 2 in the inside of front cover) and f_x is a function of the *dimensionless* quantities Z_i, $m_{\alpha i}/m_e$, ...; T_o is the atomic unit of temperature, $T_o = \hbar^2/m_e a_B^2 k_B = 315{,}775$ K, P_o is the atomic unit of pressure, $P_o = \hbar^2/m_e a_B^5 = 2.9421 \times 10^{13}$ N/m^2 $= 2.9421 \times 10^8$ bar, and $\upsilon_o = \hbar/m_e a_B = c/137 = 2187.69$ km/s is the atomic unit of velocity. The first quantity of condensed matter to be considered is the volume per atom, V/N_a, or, equivalently, the *radius per atom*, $r \equiv (3V/4\pi N_a)^{1/3}$. This radius is expected to be comparable to the atomic radius r_a but somewhat larger, because many touching spherical atoms leave always empty spaces among them, which the above definition of r incorporates; how much larger $\bar{r}$ is over $\langle \bar{r}_a \rangle$ depends mainly on the average number of nearest neighbors (which, e.g., is 12 for copper, 8 for iron, and only four for diamond). Thus

$$r = \bar{r} a_B \tag{10.2}$$

where the average value of $\bar{r}$ over all the elemental solids is 3, while the corresponding value for the atomic radius is $\langle \bar{r}_a \rangle = 2.6$. The value of $\bar{r}$ allows us to determine the density and vice versa

---

[1] A bar over any physical quantity X denotes its value at the atomic system of units: $\bar{X} \equiv X/X_o$.

$$\rho_M = m_a/(4\pi/3)r^3 = A_W m_u/(4\pi/3)a_B^3\bar{r}^3 = 2.675(A_W/\bar{r}^3)\,\text{g/cm}^3 \qquad (10.3)$$

where A_W is the average atomic weight of the condensed matter. As in the case of atoms and molecules, the replacement $a_B \to r$ in all atomic units of (10.1) is expected to improve the estimated values of various quantities X. One such quantity is the *cohesive energy* u_c defined as the minimum energy per atom required to separate the condensed matter into (neutral) atoms under conditions of $T \to 0$ and $P = 1$ atm. The latter is too small to appreciably influence u_c. Similarly, the atomic mass is expected to play a minor role, since it enters through the ionic kinetic energy which is smaller than the electronic one by a factor of $\sqrt{m_e/m_a}$, as we have seen in Chap. 9. Hence, from (10.1) we have

$$u_c = f_u(Z)(\hbar^2/m_e r^2) = f_u\hbar^2/m_e a_B^2\bar{r}^2 = f_u 27.2/\bar{r}^2\ \text{eV};\quad f_u \approx 1 \qquad (10.4)$$

The bulk modulus B at $T = 0$ is defined as $B \equiv -V(\partial P/\partial V) = V\partial^2 U/\partial V^2$. Taking into account that B has dimensions of energy over volume and using similar arguments as in the case of u_c, we obtain from (10.1) and Table 2 (inside front cover).

$$B = f_B(Z)(\hbar^2/m_e r^5) = f_B\hbar^2/m_e a_B^5\bar{r}^5 = f_B 2.94 \times 10^{13}/\bar{r}^5\ \text{N/m}^2;\quad f_B \approx 0.6 \qquad (10.5)$$

The shear modulus μ_s is of the same order of magnitude as the bulk modulus but usually smaller. For most metals, $x \equiv \mu_s/B$ is between 0.3 and 0.6, while for liquids is zero. The shear stress τ_s under which a solid fails depends mainly on linear defects called dislocations and, as a result, it varies in a very wide range of values with a gross average about equal to $\langle\tau_s\rangle \approx 0.005\mu_s \approx B/500$.

As it was mentioned in Chap. 9, the number of independent stationary ionic oscillations is huge, equal to $3N_a - 6$, and the corresponding natural frequencies cover a range from 0 to a maximum value denoted by ω_D. Connected with ω_D is the so-called Debye temperature, $\Theta_D \equiv \hbar\omega_D/k_B$, which determines the contribution of ionic oscillations to the thermodynamic properties. The quantities ω_D and Θ_D, in analogy with (8.4), are given by

$$\omega_{\max} = f_\omega\frac{e^2}{\hbar r}\sqrt{\frac{m_e}{m_a}} = f_\omega\frac{96.8}{\bar{r}\sqrt{A_w}} \times 10^{13}\ \text{rad/s};\quad \Theta_D = f_\omega\frac{7390}{\bar{r}\sqrt{A_w}}\ \text{K};\quad f_\omega \approx 1 \qquad (10.6)$$

The velocity of sound, c_o, in a liquid[2] is obtained from the unit of velocity, $\upsilon_o = \hbar/m_e a_B$, by implementing the replacement $a_B \to r$ and multiplying by the factor $\sqrt{m_e/m_a}$, since sound propagates through atomic oscillations:

[2] In isotropic solids there are two sound velocities, the transverse one, $c_t = c_o\sqrt{x}$, (in which the direction of motion of each atom is perpendicular to the direction of sound propagation), and the longitudinal one, $c_\ell = c_o\sqrt{1 + \frac{4}{3}x}$ (in which the direction of motion of each atom is along the propagation direction). $x \equiv \mu_s/B$.

Table 10.1 Comparison of the results of (10.4), (10.5), (10.7), and (10.6) with the experimental data [10.1] for four of the most important solids (technologically and historically)

	Fe		Al		Cu		Si	
	Estimate	Exp	Estimate	Exp	Estimate	Exp	Estimate	Exp
A_w		55.85		26.98		63.55		28.09
$\bar{r}$ (exp)		2.70		2.99		2.67		3.18
u_c (eV/atom)	3.73	4.28	3.04	3.39	3.82	3.49	2.69	4.63
B (10^{11} N/m^2)	1.29	1.68	0.73	0.72	1.29	1.37	0.54	0.99
c_o (km/s)	4.06	4.63	5.28	5.68	3.85	3.93	4.87	6.48
Θ_D (K)	366	464	476	426	347	344	438	645

$$c_o = f_c \frac{\hbar}{m_e r}\sqrt{\frac{m_e}{m_a}} \approx \frac{82}{\bar{r}\sqrt{A_w}}\frac{\text{km}}{\text{s}}; \quad f_c \approx 1.6 \tag{10.7}$$

Notice that (10.3), (10.5), and (10.7) together with the choices $f_B \approx 0.6$ and $f_c \approx 1.6$ lead to the simple relation $c_o = \sqrt{B/\rho_M}$, which happens to be exact (see also (5.20)). In Table 10.1 we test the results of our simple approach against experimental data.

The largest discrepancy between experimental values and our estimates appears for Si; this is due mainly to the fact that Si has only four nearest neighbors and, as a result, the ratio $\bar{r}/\bar{r}_a$ in Si is overestimated by a factor $(12/4)^{1/3} \approx 1.44$ relative to that in close-packed solids where the number of nearest neighbors is 12. If this factor of 1.44 is included, our estimates for Si will become 5.58, 3.34, 7.02, and 631, respectively. A corresponding correction for Fe involves the factor $(12/8)^{1/3} \approx 1.14$ and makes our estimates 4.88, 2.52, 4.64, 419, respectively. The above partial improvement of the estimates indicates that the atomic radius is probably a better scaling length, at least for some quantities, than the radius per atom. In other words, probably a better replacement for u_c, c_o, Θ_D is the following $a_B \to \tilde{r}a_B$, where $\tilde{r} \approx 1.1\bar{r}_a \approx \bar{r}(z/12)^{1/3}$ and z is the number of nearest neighbors (For $z = 12$ the volume fraction is 0.74; $0.74^{-1/3} = 1.1$). The results for the bulk modulus will be improved by replacing $\bar{r}^5$ by $\bar{r}^3\tilde{r}^2 = \bar{r}^5(z/12)^{2/3}$.

Next, we shall attempt to estimate the electrical resistivity, ρ, which is connected to the resistance, R, of a wire of length ℓ and cross-section s by the well known formula, $R = \rho\ell/s$. It is easy to show that the dimensions of resistance is energy × time over charge square and, hence, its unit in the atomic system is $\hbar/e^2 = 4108.236$ ohm. It follows that the unit of resistivity in the atomic system is $\rho \approx \hbar a_B/e^2 = 21.74\,\mu\text{ohm}\cdot cm$. Therefore, if we make the assumptions that T and P do not play an important role, that the atomic mass is not so important,[3] and that a_B must be replaced everywhere by $\bar{r}a_B$, we estimate that the resistivity is

[3] These assumptions seem justified if we adopt the picture that the electrical resistivity is due to the collisions of the current-carrying electrons (behaving as classical particles) with the more or less immobile ions. This picture leads to an electronic mean free path of the order of a few Angstroms.

$\rho \approx f_\rho \hbar a_B \bar{r}/e^2 = f_\rho 21.74\bar{r}$ μohm · cm, i.e., of the order of a few tens of μohm·cm. This estimate is consistent with an electronic mean free path ℓ of the order of a few Angstroms. Although there are metals with resistivities at room temperature close to this estimate (e.g., Ti with $\rho = 43.1$ μohm·cm, Pr with $\rho = 67$ μohm·cm, Mn with $\rho = 139$ μohm·cm, etc.), in general our estimate for ρ fails very badly: Clean single crystals of copper at helium temperatures have reached resistivities as low as 10^{-5} μohm·cm. To the other extreme there substances such as yellow sulfur exhibiting resistivities at room temperatures as high as 10^{23} μohm·cm.

It is very instructive to ask ourselves what went wrong in deriving an estimate of ρ which failed so dismally. As we mentioned before, our assumptions, such as weak temperature dependence, etc., rely on the picture of current-carrying electrons colliding with ions as classical particles. Hence, classical mechanics seems to be the culprit for this failure. We should have used quantum instead of classical mechanics for the motion of electrons. More specifically we should have taken into account that electrons propagate in the solid as waves, and consequently *they may exhibit constructive interference.* If the scattered electronic waves interfere constructively in *a systematic way,* the effects of scattering could be compensated and the electrons could propagate as if they were free of any force; in other words their mean free path would be infinite. However, such a systematic constructive interference requires placement of the scattering centers (i.e., the ions) in an ordered way to make sure that waves arrive in phase. The perfect crystalline structure, due to its periodic positions of the ions, satisfy this requirement and thus provides the mechanism for essentially free-like propagation, infinity mean free path, and zero electrical resistivity. According to this argument, the observed non-zero small metallic resistivity is due to elastic scattering by structural imperfections, by foreign atoms, and by other defects, as well as to inelastic scattering by the inevitable ionic oscillations (at $T \neq 0$ K). For a pure good conductor, such as noble metals and at room temperatures, the scattering by the ionic oscillations is the dominant one. This scattering is proportional to the amplitude-square of the ionic oscillations, which in turn is proportional to the energy of these oscillations. The latter at room and higher temperatures is given approximately by the classical expression of $3k_BT$ per atom. We conclude that *for good conductors and for not so low temperatures the resistivity is proportional to the absolute temperature.* Hence, we have the following expression for the resistivity of good conductors at room temperatures and higher, according to (10.1):

$$\rho_e = \text{const.}\frac{\hbar a_{\text{B}} 3T}{e^2\ T_o} \rightarrow \text{const.}\frac{3\hbar a_{\text{B}} \bar{r}}{e^2}\frac{m_e a_{\text{B}}^2 \bar{r}^2}{\hbar^2} k_B T, \quad a_B \rightarrow r = \bar{r} a_B, \tag{10.8}$$

or $\rho_e = \text{const.}\ 2.07 \times 10^{-4} \bar{r}_\alpha^3 T$ μohm · cm, for good metals (T in K and not very low).

A periodic placement of the scatterers (i.e., the ions) could lead not only to perfectly systematic constructive interference but to perfectly *systematic destructive interference* as well. So it is possible in a periodic medium to have energy regions (called *bands*) where the interference is systematically constructive and *the propagation is free-like,* and regions of energy called *gaps* where it is systematically destructive and the electronic waves cannot even exist.[4] This wave-based, alternating band/gap structure of the energy spectrum in periodic materials explains the huge differences in electrical resistivities among various solids. If in a solid there is a partially filled band at $T = 0$ K, the electrons can easily be excited by an electric field to nearby empty levels and produce thus large electric currents. Such materials behave as good conductors and their resistivity is as in (10.8). On the other hand, if every band is either fully occupied by electrons or completely empty, the resistivity (at $T = 0$ K) is infinite and the material is insulating. This is so, because the electrons in a *fully occupied band, even if they are infinitely mobile and free–like*, cannot respond to the electric field and produce a current, because any such response requires electrons to be excited from occupied to empty states. But in a *fully occupied band* there are no empty states.[5] In reality, the resistivity in insulators and semiconductors is huge but finite, because of the thermal excitation (at $T > 0$ K) of a few electrons from the highest fully occupied band, called *valence band*, to the lowest empty band, called *conduction band*.[6] The resistivity is inversely proportional to the number of carriers, i.e., the number of electrons excited in the conduction band from the valence band plus the missing electrons (*holes*) in the valence band. This number depends on the dimensionless ratio $E_g/2k_BT$, where E_g is the so-called *energy gap*, i.e., the energy separation between the bottom of the conduction band and the top of the valence band. Thus for very pure semiconductors and insulators the resistivity is proportional to $\exp[E_g/2k_BT]$

$$\rho_e \propto \exp\left[\frac{E_g}{2k_BT}\right], \quad \text{pure semiconductors, insulators.} \tag{10.9}$$

It must be stressed that the resistivity in semiconductors may be reduced by several orders of magnitudes if appropriate substitutional impurities are incorporated in the lattice, because then the gap is *effectively* reduced substantially (see Pr. 10.6).

10.2 The Jellium Model and Metals

The Jellium Model (JM) adopts the general picture that each atom participating in the formation of solids splits into ζ detached electrons and a cation of positive charge ζe. Moreover, the JM assumes that each cation is mashed as to

---

[4] It should be pointed out that gaps may appear even in non-periodic systems, if a large fraction of space is classically inaccessible to the mobile electrons. On the other hand, destructive interference in two and three dimensional periodic media is not always capable of opening gaps.

[5] There are empty states in the conduction band. However, the excitations of electrons there, requires usually a huge electric field; this is the phenomenon of electric breakdown of insulators.

[6] In insulators there may exist non-negligible electric currents due to ionic migration, especially in the presence of impurities and imperfections.

cover uniformly the volume per atom, associated with it. Thus, the JM replaces the discrete lattice of the cations by a homogeneous and isotropic continuous background of ionic mass and charge. This positive uniform ionic charge density is fully compensated by the negative charge density of the detached electrons, which, being assumed free of any force, spread themselves uniformly over the whole volume, V, of the solid. The input for any calculation based on the JM is the type (or types) of atoms making up the solid and their ζ and A_W. The drastic approximation of ignoring any force on every (detached) electron sounds less unreasonable in view of the free propagation occurring in a periodic potential. On the other hand, the JM misses completely the existence of gaps which a periodic potential produces. As a result, the JM is a model for metals and it is absolutely inappropriate for electronic transport properties of semiconductors and insulators. The starting point of any calculation within the JM is the expression of the total energy per atom, U/N_a, in terms of the radius per atom $r' = \bar{r}' a_B$:

$$\frac{U}{N_a E_o} = \frac{\alpha}{\bar{r}'^2} - \frac{\gamma}{\bar{r}'}, \quad E_o \equiv \hbar^2/m_e a_B^2 \tag{10.10}$$

The first term in the rhs of (10.10) is the kinetic energy of the detached electrons divided by N_a (see (3.9) and take into account that $V'^{2/3} \propto r'^2$), and the second is the Coulomb energy among all the charged particles (excluding the self-interaction of each cation). The values of α and γ become more realistic than those resulting from the JM, if we reintroduce the cations as spherical particles of radius $r_c = \bar{r}_c a_B$, while still considering the detached electrons as uniformly spread in the remaining volume $V' = V - N_a(4\pi/3)r_c^3$. We have then $\alpha = 1.3\zeta^{5/3} + \eta \bar{r}_c \zeta^2$ and $\gamma = 0.56\zeta^{4/3} + 0.9\zeta^2$ where η varies between 0.4 and 0.9 (the lower value is for ions of half-filled subshells and the highest for ions of completed subshells). (see [10.1])

By minimizing U/N_a, with respect to r' we obtain its equilibrium value $\bar{r} = 2\,\alpha/\gamma$. For the solids Fe, Al, Cu, and Si the results for $\bar{r}$ are 2.54, 3.01, 2.74, 2.733, respectively, to be compared with the experimental values given in Table 10.1. Using the definition of B (see line below (10.4)), and (10.10) we find that, $B = (\alpha/6\pi\bar{r}^5)P_o$, $P_o \equiv \hbar^2/m_e a_B^5$, under zero pressure and temperature. The results for the four solids of Table 10.1 are 1.97, 0.85, 1.04, 2.51, respectively. Results for the sound velocity, $c_o = 24.15\sqrt{\alpha}/\bar{r}\sqrt{A_W}\,\text{km/s}$, and the Debye temperature, $\Theta_D = 8426(\alpha/A_W)^{1/2}(f/\bar{r}^2)$ K, can be obtained as well. The value of f depends on the ratio $x = \mu_s/B$ and is found to be 0.78, 0.64, 0.67, 0.913 for the Fe, Al, Cu, and Si, respectively. Within the JM the stationary oscillations of the ions are plane acoustic waves of frequency $\omega = 2\pi c_o/\lambda$ (for liquids, see (Pr. 10.5), while for solids we have transverse and longitudinal acoustic waves (see footnote 2).

The JM supplemented by some scattering processes let us calculate the response of a metal to external stimuli, such as an electric field, and/or a magnetic

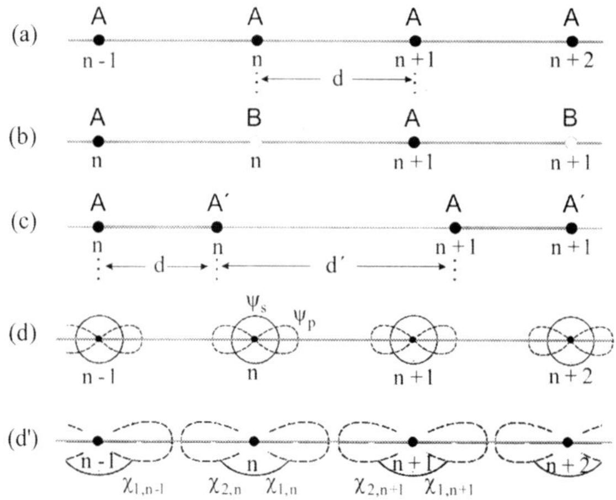

Fig. 10.1 **a** An elemental 1D periodic metallic "solid", **b** an ionic 1D "solid" consisting of two kinds of atoms; the period is $a = 2d$, **c** a molecular 1D "solid"; the period is $a = d + d'$. **d** an elemental periodic 1D "semiconductor" with two relevant atomic orbitals and two electrons per atom. (**d′**) As in (**d**) but with the χ_1, χ_2 hybrids instead of the atomic orbitals ψ_s, ψ_p

field and/or a temperature gradient [10.1]. This response is characterized in the linear regime by quantities such as the resistivity, the dielectric function, the magnetic permeability, the thermal conductivity, etc.

10.3 The LCAO Method and Semiconductors

The LCAO method let us study solids such as semiconductors, ionic solids, and molecular solids where the JM is not adequate. In Fig. 10.1 we plot schematically various types of one-dimensional[7] "solids": In cases (a), (b), and (c) each atom A or B has one electron in an atomic orbital ψ_A or ψ_B with corresponding energies ε_A or ε_B, respectively; In case (d) there is one electron in each of the two relevant atomic orbitals s and p_x. There are also non-zero off-diagonal matrix elements V_2 which can transfer the electron from one atom to each of the two nearest neighbors. The stationary electronic waves for the periodic systems of Fig. 10.1, called here Bloch-states (B-states), and the corresponding energies are as follows (see [10.1]):

Case (a) The B-states are as in (9.15), with 6 replaced by the number of atoms N_a. The energy of each B-state is $\varepsilon_\nu = \varepsilon_A + 2V_2 \cos \varphi_\nu$. Two successive $\varphi_\nu s$ differ by $2\pi/N_a \to 0$, as $N_a \to \infty$. Hence, the energy spectrum becomes dense giving rise to a band of total energy extent equal to $4|V_2|$. The number of B-states in this band are N_a. Placing two electrons in each of the lower B-states we end up with a half-filled band and, hence, with a conducting, metallic behavior.

[7] One-dimensional "solids" allow us to present basic concepts in the simplest possible way.

Case (b) The B-states now are of the form, $\phi = \sum_n c_{n,A}\psi_{n,A} + \sum_n c_{n,B}\psi_{n,B}$. The coefficients, $c_{n,A}$, $c_{n,B}$ satisfy the relation $c_{n,i} = c_i \exp(in\varphi)$, $i = A, B$, as a result of Bloch's theorem stating that a translation by the period a multiplies the state by a phase factor $\exp(i\varphi)$. We define the so-called crystal wavenumber $k \equiv \varphi/a$. Its values range from $-\pi/a$ to π/a in steps of $2\pi/(N_a/2)a = 2\pi/L$. The energy of the B-state can be found as in (9.7). By minimizing this energy with respect to c_A, c_B, we find a system of two equations for these unknown coefficients,

$$(\varepsilon_A - \varepsilon_k)c_A + V_2\left(e^{-\mathrm{ika}} + 1\right)c_B = 0, \quad (\varepsilon_B - \varepsilon_k)c_B + V_2\left(1 + e^{\mathrm{ika}}\right)c_A = 0 \quad (10.11)$$

This system has non-zero solutions only if the energyε_k satisfies the relations

$$\varepsilon_\pm(k) = \varepsilon \pm \sqrt{V_3^2 + 4V_2^2\cos^2(\mathrm{ka}/2)} \quad (10.12)$$

where ε and V_3 are defined as in (9.9). Obviously (10.12) implies that there are two bands, the lower one extending form $\varepsilon - \sqrt{V_3^2 + 4V_2^2}$ to $\varepsilon - V_3 = \varepsilon_B$ and the upper one from $\varepsilon + V_3 = \varepsilon_A$ until $\varepsilon + \sqrt{V_3^2 + 4V_2^2}$with a gap equal to $E_g = \varepsilon_A - \varepsilon_B$. Each band has $N_a/2$ B-states. Hence, at $T = 0$ the lower band is fully occupied and the upper one is empty. Therefore, we have the case of an ionic insulator, e.g. NaCl.

Case(c) Following a similar procedure as in case (b) we find for the energies

$$\varepsilon_\pm(k) = \varepsilon_A \pm \sqrt{V_2^2 + V_2'^2 + 2V_2V_2' \cos \mathrm{ka}} \quad (10.13)$$

where $V_2 \propto d^{-2}$, $V_2' \propto d'^{-2}$ are the large and the small matrix elements, respectively. Again we have two bands, the lower fully occupied and the upper empty; the gap is $E_g = 2\left|V_2 - V_2'\right|$. This is the case of a molecular insulator, such as solid N_2.

Case (d) This case can be analyzed by employing hybrids as shown in Fig.10.1d′. Two neighbouring hybrids, e.g., $\chi_{1.n}$ and $\chi_{2.n+1}$, form a molecular bonding and a molecular antibonding level at each bond as in diatomic molecules. Because of the non-zero matrix element V_1 connecting $\chi_{1.n}$ and $\chi_{2.n}$, each electron in the bonding molecular orbital propagates from bond to neighbouring bond. This propagation is equivalent to that in case (a) with ε_b playing the role of ε_A and $V_1/2 = (\varepsilon_s - \varepsilon_p)/4$ playing approximately the role of V_2. Hence, a band is create around the bonding level ε_b of total width approximately equal to $(\varepsilon_p - \varepsilon_s)$. Similarly a band of approximate width $(\varepsilon_p - \varepsilon_s)$ is formed around the antibonding molecular level ε_a. It follows that the gap is (see also [10.1], Chap. 7, and Pr. 10.3):

$$E_g \approx (\varepsilon_a - \varepsilon_b) - (\varepsilon_p - \varepsilon_s) = 2|V_{2h}| - (\varepsilon_p - \varepsilon_s) = 6.4\frac{\hbar^2}{m_e d^2} - (\varepsilon_p - \varepsilon_s) \quad (10.14)$$

Each of the bonding and antibonding bands has N_a states. Hence, the bonding band can accommodate exactly all the electrons, becoming thus the valence band, while the antibonding band is empty at $T = 0$ being so the conduction band. This is typical case of semiconductors. The gap, according to (10.14), for the elemental solids of the carbon column starting with the diamond, Si, Ge, Sn, and Pb, is 12.28, 1.66, 0.42, -0.02, and -1.95 eV, respectively, while the corresponding experimental values are 5.5, 1.17, 0.74, 0, metal. A negative value of the gap indicates overlap of the two bands and, hence, a metallic behaviour. The tendency for the gap to be reduced as we go down the carbon column is mainly due to the systematic increase of the size of the corresponding atoms and the resulting increase of the bond length, which implies a smaller $|V_{2h}|$ and a smaller gap.

Real semiconductors in 3D employ the $sp^{(3)}$ hybridization and, hence, are arranged in space in a tetrahedral way with four nearest neighbours.

Part IV
Gravity at Front Stage

When is the gravitational energy comparable with the Coulomb energy in a neutral system?

The gravitational energy U_G of an electrically neutral system of mass M grows with M as $U_G = aM^{5/3}$ (assuming that the density does not depend on M), while the Coulomb energy grows linearly with M as $U_C = bM$; the quantities a and b will be examined below. This difference in the rate of growing is due to the long range and always attractive character of the gravitational interaction versus the *effectively* short range character of the Coulomb interaction in an overall electrically neutral system. This effective short range property is due to the Coulomb interaction being either attractive, for opposite charged particles, or repulsive, for particles of the same charge. As a result in overall neutral systems positively and negatively charged particles combine to form *locally* neutral entities which attract each other through the short range residual van der Waals interaction. The gravitational energy will become equal to the Coulomb energy when the total mass will be equal to

$$M = (b/a)^{3/2} \tag{IV.1}$$

This characteristic exponent of 3/2 will appear very often in calculation involving gravity.

The gravitational self energy of a spherical body of mass M and radius R, according to Newton's law, equals to

$$U_G = -\frac{3x}{5}\frac{GM^2}{R} \tag{IV.2}$$

where x equals 1, if the mass density is homogeneous and isotropic; if the density is decreasing as the distance from the center is increasing, x is larger than 1. (For Earth, $x \approx 1.025$; see (11.11)). Taking into account that $M = (4\pi/3)R^3 \rho_M$ and (9.3) we obtain that

$$a = -\frac{3x}{5}\frac{GA_W^{1/3} m_u^{1/3}}{a_B \bar{r}} \tag{IV.3}$$

A_w is the average atomic weight.

The Coulomb energy was obtained before in (10.10). Substituting in this formula, N_a from the relation $M = m_u A_W N_a = m_u N_\nu$, where N_ν is the total number of nucleons, we find for b

$$b = -\frac{\gamma y}{\bar{r}} \frac{e^2}{a_B} \frac{1}{A_W m_u} \tag{IV.4}$$

where y is a correction numerical factor to take into account a non-uniform concentration of particles; y is expected to be a little larger than 1, but probably smaller than x. Combining (IV.1), (IV.3), (IV.4), and, $M = m_u N_\nu$, we find that, in an electrically neutral system, the number of nucleons N_ν, which makes the gravitational energy equal to the Coulomb energy:

$$N_\nu = \left(\frac{5y}{3x}\right)^{3/2} \frac{\left(0.56\zeta^{4/3} + 0.9\zeta^2\right)^{3/2}}{A_W^2} \left(\frac{e^2}{G m_u^2}\right)^{3/2} \tag{IV.5}$$

where γ was set equal to $0.56\zeta^{4/3} + 0.9\zeta^2$ (see p. 85). The dimensionless ratio $e^2/Gm_u^2 \approx 1.2536 \times 10^{36}$ is the ratio of the strengths of the electromagnetic to the gravitational interactions (see Table 2.2, p. 10). The radius R corresponding to N_ν is given by

$$R = \bar{r} a_B N_a^{1/3} = \bar{r} a_B N_a^{1/3} / A_W^{1/3}$$

If we choose $A_W \approx 40$ (as in Earth), $\zeta = 1$, $y/x = 1$, and $\bar{r} = 2.75$, we find $N_\nu = 3.3 \times 10^{51}$ and $R = 6.35 \times 10^6$ m versus $N_\nu = 3.596 \times 10^{51}$ and $R = 6.371 \times 10^6$ m for Earth. If we choose $A_W \approx 2, \zeta \approx 1, y/x \approx 1, \bar{r} \approx 1.6$, values close to those of Jupiter, we obtain $N_\nu \approx 1.33 \times 10^{54}$ and $R \approx 7.36 \times 10^7$ m versus $N_\nu \approx 1.143 \times 10^{54}$ and $R \approx 7.15 \times 10^7$ m for Jupiter. We conclude that for the Earth and for Jupiter the Coulomb energy is comparable to the gravitational energy. For the Sun the ratio U_G/U_C (assuming that its and A_W and ζ are as those of Jupiter) is equal to

$$U_G/U_C \approx (a/b) M_s^{2/3} \approx (M_S/M_J)^{2/3} \approx 100 \tag{IV.6}$$

where M_S, M_J are the masses of the Sun and the Jupiter respectively.

Chapter 11
Planets

Abstract In planets the gravitational potential energy is comparable to the electrostatic potential energy. The central star sends energy and information to the planets around it determining thus their temperature and some of the other conditions necessary for the rocky planets to be candidate cradles of life.

Keywords Mountains · Temperature · Winds · Pressure

Summary In this chapter we shall examine a few properties pertaining to planets. First, why planets are of spherical shape? Yet, why are there mountains in rocky planets? We find that the maximum possible height of a mountain in a rocky planet is determined by the competition between the weight of the mountain and the mechanical failure of the rocky material. Second, what determines the average temperature of a planet? The answer is: The balance between the E/M energy received by a planet from the central star and the E/M energy emitted by the planet. We show that the temperature difference between the planet and the central star leads to a net emission of entropy by the planet, a fact that makes the phenomenon of life possible: Rocky planets are candidate cradles of life. Finally, we briefly examine the question of why do winds appear in the atmospheres of planets? Other information and data regarding planets can be found in (11.1) and (11.2).

11.1 How Tall Can a Mountain Be?

Planets and other celestial large bodies are of spherical shape, because this shape is the one which minimizes the total gravitational energy (assuming isotropy in the chemical composition). This seems rather obvious, since any other shape can come from the spherical one by removing appropriate pieces of matter and placing them at a larger distance from the center, increasing thus the energy. Hence, the mountains or ridges in a rocky planet, since they represent deviations from the

E. N. Economou, *A Short Journey from Quarks to the Universe*,
SpringerBriefs in Physics, 1, DOI: 10.1007/978-3-642-20089-2_11,

spherical symmetry, increase its total energy. Mountains (and ridges) were created during the formation of a planet and/or during major "geological" activities over the ages. What keeps them existing is an energy barrier that does not allow the spontaneous flow of matter, which would lead to a flattened surface. This barrier involves the deformation and eventual breaking of chemical bonds, which macroscopically lead to failure of the solid state either by plastic flow or by propagation of cracks. This failure will occur when the *shear* stress, τ, at the base of a mountain is equal to its critical value τ_s. The latter has been estimated in p. 83 to be of the order of $B/500$. The shear stress at the base of a mountain is of the order of the pressure $P = W/S$, where W is the weight of the mountain and S is its cross-section at its base. Assuming a conical shape of the mountain, we get $W = \frac{1}{3} S H \rho_M g$, where H is the height of the mountain. Hence, by equating $\tau = \tau_s$ we obtain the criterion determining the highest possible mountain in a rocky planet:

$$\frac{1}{3} x H \rho_M g = y \frac{0.6}{500} \frac{\hbar^2}{m_e a_B^5 \bar{r}^5} \tag{11.1}$$

where x, y are numerical factors of the order of unity. We shall take as rough estimate $x/y = 1$. The density has been calculated in Chap. 10: $\rho_M = A_W m_u / (4\pi a_B^3 \bar{r}^3/3)$. The acceleration of gravity g is given by $g = GM/R^2 = G(4\pi R^3/3)\rho'_M/R^2 = \frac{4\pi}{3} G R \rho'_M$, where the prime indicates that we refer to the whole planet, while the unprimed quantities refer to the crust. Substituting in (11.1) we obtain

$$\frac{HR}{a_B^2} = 0.015 \frac{y}{x} \frac{\bar{r}'^3}{\bar{r}^2} \frac{1}{A_W A'_W} \frac{e^2}{G m_u^2} \tag{11.2}$$

Choosing the Earth's values, $\bar{r} = 2.89$, $\bar{r}' = 2.69$, $A_W = 24.25$, $A'_W = 40$, we find $HR \approx 10^{11}\ \text{m}^2$ or $H \approx 16$ km. This means that it is not possible for a mountain on Earth to be taller than about 16 km. For Mars with $R = 3396$ km and the product $HR \approx 1.97 \times 10^{11}\ \text{m}^2$ (assuming that $\bar{r}$, $\bar{r}'$ are the same as in Earth and that $(A_W A'_W)_E/(A_W A'_W)_M \approx (\rho_{M,E}/\rho_{M,M})^2$) the largest possible height of a mountain cannot exceed 59 km. The highest mountain in Mars, Mt. Olympus, is about 24 km [11.1].

A rocky asteroid may have a potato-like shape, which implies that $H \approx R$. Combining this last relation with $HR \approx 2 \times 10^{11}\ \text{m}^2$ (as in Mars), we obtain a rough estimate of the biggest asteroid having such a potato-like shape: its extent is about 500 km.

11.2 Temperature of a Planet

The surface[1] temperature of a planet is determined by equating the absorbed E/M power, I_a, with the emitted one, I_e; the latter, according to (5.3), is given by $I_e = 4\pi R_P^2 \sigma T_P^4$; $\sigma \equiv \pi^2 k_B^4 / 60\hbar^3 c^2$ is the Stefan–Boltzmann constant, The power,

[1] The surface includes also the main volume of the atmosphere of the planet.

I, emitted by the central star, equals $I = 4\pi R_S^2 \sigma T_S^4$; this power passes through the surface of a sphere of radius R equal to the mean distance between star/planet. The planet intersects a small fraction of I equal to $\pi R_P^2/4\pi R^2$. Hence, the power it absorbs is $I_a = (1 - A) I \pi R_P^2 / 4\pi R^2 = (1 - A)(4\pi\sigma R_S^2 T_S^4)\pi R_P^2/4\pi R^2$, where A is the reflection coefficient of the planet's surface, the so-called albedo (for Earth $A \approx 0.3$). Substituting the above relations in the equality $I_a = I_e$ we obtain the planet's surface temperature T_P:

$$T_P = (1 - A)^{1/4} (R_S/2R)^{1/2} T_S \tag{11.3}$$

The ratio R_S/R is equal to $\tan(\theta/2) \approx \theta/2$, where θ is the viewing angle of the star from the planet. For the Earth/Sun system, $\theta = (32/60)(2\pi/360) = 0.0093$ and the Sun's surface temperature is 5800 K. Hence, the surface temperature of Earth is, according to (11.3), 256 K, very close to the accepted value of 254 K.

Actually, for Earth, about 25% of I_e is emitted from the ground at a temperature T_g, another 25% is emitted by the clouds at about 3 km with a temperature of T_c, and the rest of 50% is emitted from the top of the troposphere at about 10 km and with a temperature T_t. Thus

$$I_e = 4\pi R_P^2 \sigma T_P^4 = 4\pi R_P^2 \sigma \left(\frac{1}{4} T_g^4 + \frac{1}{4} T_c^4 + \frac{1}{2} T_t^4 \right) \tag{11.4}$$

By assuming mechanical equilibrium in the atmosphere and constant entropy as a function of height, the temperatures T_c, T_t can be expressed in terms of T_g as follows: $T_c = T_g - 25$ K, $T_t = T_g - 65$ K. Substituting these relations in (11.4) we find that $T_g \approx 290$ K, $T_c \approx 265$ K, $T_t \approx 225$ K. These numbers are very important for the appearance and continued existence of life on Earth, since $T_g \approx 290$ K keeps water in a liquid form on Earth's ground.

It is worth to point out that the flow of E/M radiation in and out of a planet implies an external change in its entropy given by $đS_e/dt = (I_a/T_S) - (I_e/T_P)$, which is clearly negative due to the fact that $T_P \ll T_S$. Consider now an initial state of the planet such that $đS_i/dt$ is smaller than $|đS_e/dt|$; then, the total entropy of the planet will be decreasing which implies increasing organization possibly through new structures. The latter ought to be of dynamical rather than of static nature for the following reason: These new structures have to *continuously* produce entropy as to match its external reduction and to reach eventually a steady state such that $dS/dt = 0$. In conclusion, the in and out flow of E/M radiation in a planet leads to a decrease of its entropy, which in turn may set up ordered dynamical structures continuously producing an internal increase of its entropy up to a rate of fully compensating the external one. Such a dynamical structure of extreme complexity, extent, and, possibly, rarity is the living matter on Earth with its continuous evolution. Thus life may exist in open systems, such as planets, not in spite of, but because of the Second Law of Thermodynamics. Of course, the appearance of living matter requires many other conditions to be satisfied; very few of them are known with certainty.

11.3 Winds in the Atmospheres of Planets

Winds in the atmospheres of planets appear mainly because of the non-uniform absorption of the E/M radiation emitted by the central star. This non-uniformity is due to the rotation of the planet around its axis and the resulting day/night temperature difference, as well as to the fact that the equatorial region absorbs more radiation than the poles. The average of the square of the velocity of winds, υ^2, is directly connected with the average kinetic energy, E_K, of the atmosphere. This kinetic energy depends obviously on the E/M power absorbed, I_a, since this is the only source of energy. For dimensional reasons we need a characteristic time, τ, by which to multiply I_a to get E_K:

$$E_K = \eta I_a \tau \tag{11.5}$$

where $E_K = \frac{1}{2} M_a \upsilon^2$, M_a is the total mass of the atmosphere, and η is a numerical factor of the order of one. What is the characteristic time τ? It turns out that it is the time required for a sound wave to go around half the circumference of the planet, i.e. the shortest time required for communication between the night region of minimum absorption to the day region of maximum absorption: $\tau = \pi R_P / c_o$; the velocity of sound, c_o, according to (10.7) and the text in p. 82, is given by $c_o = \sqrt{B/\rho_M}$, where $B \equiv -V(\partial P/\partial V)$. The derivative in the definition of B is to be taken under constant entropy, since during the sound oscillations there is no time either for exchange of heat or for appreciable entropy production. If the atmosphere is assumed to be a perfect gas obeying the equation of state $PV = Nk_BT$, we have that $V(\partial P/\partial V)_T = -P$ which combined with the general relation $(\partial P/\partial V)_S = (C_P/C_V)(\partial P/\partial V)_T$ gives $B_S \equiv -V(\partial P/\partial V)_S = (C_P/C_V)P$. The density $\rho_M = NA_W m_u/V$, where A_w is the average atomic (or molecular) weight ($A_W = 28.96$ for Earth's atmosphere). Substituting in the expression for c_o, we have

$$c_o = \sqrt{\frac{C_P}{C_V} \frac{RT}{A_W \times 1g}} \tag{11.6}$$

where $R \equiv N_A k_B = 8.31$ J/mol·K is the gas constant. Setting $\tau = \pi R_P / c_o$ in (11.5) we end up with the following rather complicated formula, which was first obtained empirically:

$$\sqrt{\upsilon^2} = a\, \sigma^{1/16} q^{7/16} c_P^{-1/4} R_P^{1/2} \mu^{-1/2} \tag{11.7}$$

where $a = \sqrt{\eta/2}$ (≈ 0.6 for Earth), $\sigma \equiv \pi^2 k_B^4 / 60\hbar^3 c^2$, $q = (1 - \frac{4}{4})(I_S/4\pi R^2)$, R is the distance planet/star, $c_P = C_P/M_a$ is the specific heat per unit of mass of the atmosphere under constant pressure, and $\mu = M_a/4\pi R_P^2$.

11.4 Pressure and Other Quantities Within a Planet

The dependence of the mass density of a planet as a function of the distance r from the center will be approximated as follows:

$$\rho_M = \rho_o - a\,r, \quad a = (\rho_o - \rho)/R_P \tag{11.8}$$

where ρ is the mass density at $r = R_P$. Then the mean density is $\bar{\rho} = \rho_o - 3\,a\,R_P/4 = \frac{1}{4}\rho_o + \frac{3}{4}\rho$. The acceleration of gravity at r is equal to $g(r) = -GM(r)/r^2$, where $M(r)$ is the mass within a sphere of radius r. The final result is

$$g(r) = -\frac{4\pi}{3}G\,r\left(\rho_o - \frac{3}{4}(\rho_o - \rho)\frac{r}{R_P}\right) \tag{11.9}$$

By integrating $g(r)$ over r and determining the integration constant from the value at $r = R_P$ we obtain the potential $\mathsf{V}(r))$:

$$\mathsf{V}(r)) = \frac{2\pi}{3}G(\rho_o r^2 - \frac{1}{2}a\,r^3) + C,\ C = -(3\,G\,M/2R_P) + \frac{\pi}{3}G\,a\,R_P^3 \tag{11.10}$$

where $M \equiv M(R_P)$. By integrating $\mathsf{V}(r))\rho(r)d^3r/2$ we can calculate the gravitational self-energy U_G:

$$U_G = -\frac{3}{5}\frac{G\,M^2}{R_P}\left(1 + \frac{5}{48}\frac{\rho_o - \rho}{\rho_o} - \frac{5}{56}\left(\frac{\rho_o - \rho}{\rho_o}\right)^2\right) \tag{11.11}$$

Finally, the pressure at r can be obtained as the weight per unit cross-section of a *cylindrical* column along the radial direction extending from r to R_P:

$$P(r) = \int_r^{R_P} g(r')\rho(r')dr' \tag{11.12}$$

The pressure at the center is

$$P(0) = \frac{1}{2}\frac{GM}{R_P^2}\bar{\rho}\,R_P\,F = \frac{F}{2}g(R_P)\,\bar{\rho}\,R_P \tag{11.13}$$

where $F = (\rho_o/\bar{\rho})^2 - \frac{7}{6}[\rho_o(\rho_o - \rho)/\bar{\rho}^2] + \frac{3}{8}[(\rho_o - \rho)^2/\bar{\rho}^2]$.

For Earth we have that $\bar{\rho} = 5.515\,\text{g/cm}^3$, $R_P = 6371$ km; we choose approximately $\rho_o = 14$ g/cm^3 and $\rho = 2.69$ g/cm^3 in order to reproduce correctly the average density and to reasonably represent the actual complicated $\rho(r)$. We find for the pressure at the center of the Earth $P(0) = 3.44$ Mbar versus 3.51 Mbar for the actual value.

Chapter 12
Stars Dead Or Alive

Abstract In active stars the gravitational compressive pressure is balanced by the pressure due to the thermal motion of particles and to that of photons. This balance is maintained as long as the fusion nuclear reactions in the interior of the star take place. When the nuclear fuel is exhausted the mighty gravity compresses the star to a black hole, unless the pressure of the quantum kinetic energy of either the electrons (case of white dwarfs) or the neutrons (case of pulsars) is enough to balance the gravitational pressure. The lower limit of the mass of an active star is about 0.08 times the mass of the Sun. The upper limit is probably of the order of 100 times the mass of the Sun.

Keywords Mass limits · White dwarfs · Neutron stars · Black holes

Summary In this chapter we examine the limits within which the mass of an active star ranges and what determines these limits. An active star is a stellar object whose almost steady state is maintained by hydrostatic equilibrium between the gravitational pressure on the one hand, and the radiation pressure and the pressure due to thermal kinetic energy on the other; the latter pressures are fed by the fusion nuclear reactions taking place within the star. When the nuclear fuel is exhausted and after a relatively short period of instability, the star reaches a true thermodynamic equilibrium state where the gravitational squeeze is balanced, as we shall show, by the quantum kinetic energy of either the electrons (in a white dwarf) or mainly the neutrons (in a neutron stars). Pulsars, being rotating neutron stars, present the most extreme known state of matter and apparently the most effective in resisting the gravitational squeeze. However, if the mass exceeds a critical value, even this extreme state of matter is unable to withstand gravity. If this happens, the result seems to be a collapse to a black hole.

E. N. Economou, *A Short Journey from Quarks to the Universe*,
SpringerBriefs in Physics, 1, DOI: 10.1007/978-3-642-20089-2_12,

12.1 Introduction

The process of star formation occurs in gaseous interstellar media, more specifically in regions of increased density called *molecular clouds*. The largest among them, called giant molecular clouds, have concentrations of about 100 particles per cm^3 (versus 0.1–1 particles per cm^3 for the average interstellar medium), linear extent of about 100 light years, and total mass of about 6×10^6 solar masses. For a number of reasons, material consisting mainly of hydrogen molecules and helium atoms start to coalesce towards some points within the cloud, thus setting up in motion the process of the parallel formation of several stars. The resulting higher local density around these points attracts more material which arrive with ever increasing kinetic energy as a result of the growing attractive gravitational force. Collisions among the coalescing particles tends to restore a temporary equilibrium which keeps evolving towards higher temperatures and concentrations as the gravity continues to bring in more particles of higher kinetic energy. Thus a gaseous spherical body of radius R, temperature T, and mass M is formed. If the final value of mass M, reached when the surrounding material is exhausted, exceeds a critical value M_1, the temperature is high enough to ignite fusion nuclear reactions, and to turn the body to an active star. On the other hand, as we shall show in Sect. 12.6, if the final mass is too high, the radiation pressure will blow the star apart.

In an active star the fusion reactions supply enough energy to substitute for the emitted energy (mainly in the form of E/M radiation) and to maintain the kinetic and radiation pressure required to counterbalance the gravitational one. Details regarding star formation and their dynamical equilibrium between the fusion nuclear reactions and the gravitational squeeze can be found in several excellent textbooks dealing with the fascinating and growing subject of astrophysics, see for example references [11.2 and 12.1].

When the nuclear fuel within the star is exhausted, the star enters in a relatively short period of instability during which it looses an appreciable percentage of its mass, either through the so-called stellar wind,[1] if the mass is small, as in the case of our sun, or, mainly, through an explosive process known as supernova, if the mass is large (between, roughly, 10–20 solar masses). The remaining mass M', after the end of the instability period, has at its disposal only the quantum kinetic energy to counterbalance the gravitational squeezing pressure.[2] As it will become evident below, the first stage of halting the gravitational collapse is through the quantum kinetic energy of the electrons, which have been detached from the atoms. This state of matter resembles the metallic one examined in Sect. 10.2. Indeed, in metals, the quantum kinetic energy of the detached (valence only)

---

[1] Stellar winds, powered by radiation pressure, occur during the whole life of an active star although at a low rate; this rate is greatly accelerated during the period of instability.

[2] It is assumed here that the mass M' is large enough, e.g., larger than $10^{55} m_u$, so that the gravitational energy dominates over the Coulomb energy.

electrons balance the Coulomb energy. In the present case, more detached electrons are needed (usually all of them) to balance the much stronger gravitational energy. Thus, the first resistance front against the gravitational collapse of a dead star involves a fully ionized spherical body, called *white dwarf*, consisting of electrons and naked nuclei. This kind of structure is effective up to a mass M_1' . If the mass[3] M' exceeds M_1', the gravitational pressure will squeeze the dead star, until the second and probably the last resistance front is formed. This second front involves a structure called *neutron star* or *pulsar* in which all kind of particles participate: neutrons, protons, and electrons. Pulsars are successful in halting the gravitational collapse up to a mass M_2'; beyond this limiting value, no structure seems to exist capable of stopping the gravitational squeeze, which continues without limit giving rise to what is called a *black hole*.

12.2 White Dwarfs

Let us assume that the quantum kinetic energy of the electrons within a white dwarf has not approached the relativistic limit. Then the total kinetic energy is $E_K = 2.87\hbar^2 N_e^{5/3}/m_e V^{2/3} = 2.87(3/4\pi)^{2/3}\hbar^2 N_e^{5/3}/m_e R^2 = 1.1\hbar^2 N_e^{5/3}/m_e R^2$ (see (3.9)), where R is the radius of the white dwarf. The numerical factor 1.1 is valid for a system of uniform density. For a white dwarf one can calculate the density as a function of the distance r from its center [4.2] and as a result will find that 1.1 is replaced by a larger constant denoted here by b. The gravitational energy E_G is given by (3.3.2) with the constant $3x/5$ denoted by a. Minimizing the sum $E_K + E_G$ with respect to the radius R we find

$$\frac{R}{a_B} = \frac{2b}{a}\left(\frac{N_e}{N_v}\right)^{5/3}\frac{e^2}{Gm_u^2}\frac{1}{N_v^{1/3}} \tag{12.1}$$

The ratio $2b/a$ is equal to 4.51 instead of the value 2.2/0.6 = 3.68 for a uniform system. The ratio N_e/N_v is just a little lower than 0.5, if we assume that all the hydrogen has been burnt during the active life of the star; we shall choose the value 0.5 for this ratio. The universal constant e^2/Gm_u^2 is equal to 1.2536×10^{36}. Taking for the number of nucleons in the white dwarf the typical value of about half the present number of nucleons in the sun, $\frac{1}{2}N_{v,S} = 0.5986 \times 10^{57}$, we find that $R = 1.11 \times 10^7$ m, i.e., about 75% larger than the radius of the Earth.

The kinetic energy per electron $\varepsilon_k \equiv E_K/N_e$, taking into account (12.1), is proportional to $N_v^{4/3}$. Hence, for large enough N_v the electron velocities will be relativistic and will eventually reach the extreme relativistic limit for which (3.7) is valid. Implementing in (3.7) the exclusion principle replacement of $V^{1/3}$ by

[3] It is not unusual for a dead star to suck up mass from a companion star so that its mass M' to be larger than its remaining mass just after the end of its active cycle.

$(2V/N_e)^{1/3} \propto R/N_e^{1/3}$ we have for the total quantum kinetic energy $E_K = b_1 \hbar c N_e^{4/3}/R$, where the constant b_1 is 1.44 for a uniform system and higher for the actual density distribution in a relativistic white dwarf.

Notice that, in the extreme relativistic limit, the total kinetic energy becomes B/R, i.e., of the same R dependence as the gravitational energy which, of course, is of the form $-A/R$. Hence, if $B > A$, the minimization implies that R will increase until we return to the non-relativistic limit, while, if $B < A$, R will decrease until the neutron star structure is reached. Therefore, the critical value M_1' for the collapse of the white dwarf is determined by the relation $A = B$, which gives

$$N_{\nu 1}' = \left(\frac{b_1}{a}\right)^{3/2} \left(\frac{N_e}{N_\nu}\right)^2 \left(\frac{\hbar c}{G m_u^2}\right)^{3/2} \tag{12.2}$$

The ratio $b_1/a = 2.126$ instead of $1.44/0.6 = 2.4$ for a uniform system; for the ratio of the number of electrons over the number of nucleons we take the same number as before, 0.5. The universal constants $\hbar c/G m_u^2 = 1.717886 \times 10^{38}$. Replacing in (12.2) we find $N_{\nu_1}' = 1.7179 \times 10^{57}$ or $N_{\nu_1}' \big/ N_{\nu,s} = 1.43$. The conclusion is that no white dwarf with mass greater than about 1.4 the mass of the sun can exist. This agrees with all the observational data and it is known as the *Chandrasekhar limit*. Dead stars with mass $M' > M_1' = 1.4 M_s$ will be either neutron stars or black holes, although the hypothetical existence of a *quark star* has been examined theoretically as a state of matter after the gravitational collapse of a neutron star and before the ultimate collapse to a black hole.

12.3 Neutron Stars or Pulsars (≡Rotating Neutron Stars)

The analysis of the previous section implies that for $M' > M_1'$ the quantum kinetic energy of the electrons cannot stabilize the structure no matter how small the radius R can become. Thus the only possibility for equilibrium is through the contribution of the non-relativistic kinetic energy of either the nuclei or the nucleons, depending on the equilibrium value of R. If R becomes so small that nuclei will be squeezed partly inside each other, their individuality will be lost and the whole star will become a super huge nucleus held together by the mighty gravitational pressure. To see if this is the case we shall start with the opposite hypothesis, that the nuclei, by retaining their individual existence, produce an equilibrium value of R which does not imply squeezing the nuclei inside each other. Under this hypothesis the non-relativistic quantum kinetic energy of the nuclei must balance a fraction of the gravitational energy, for example half of it. The result for R would be as in (12.1) multiplied by $2(m_e/m_a) = 2(m_e/m_u A_W)$ and divided by $Z^{5/3}$. This implies a reduction of (12.1) by a factor of the order of 10^{-6}, which means a radius R of about 10 m, and a nearest neighbor distance among nuclei of the order of 10^{-2} fm! This is obviously inconsistent with the hypothesis that individual nuclei will be able to stabilize the dead star; hence, stabilization will be achieved, if at all, by the nuclei being dissolved into a closed packed "soup" of nucleons including the electrons. Moreover, the electrons will react with protons producing neutrons and vice versa.[4] When this reaction reaches equilibrium, the Gibbs free energy becomes minimum which implies the following relation among the chemical potentials of neutrons, protons, and electrons: $\mu_n = \mu_p + \mu_e$. Using this relation in combination with the equality $E_K = -\frac{1}{2}E_G$ and taking into account that $\mu_i = \partial E_{Ki}/\partial N_i$, $i = n, p, e$, we find that the numbers of neutrons, protons, and electrons at equilibrium can be expressed in terms of the number of nucleons as follows: $N_n^{'} = xN_\nu, N_p^{'} = N_e^{'} = yN_\nu$, where $x \approx 0.934$ and $y \approx 0.066$. In obtaining these relations we assumed that the neutrons and protons are non-relativistic, in contrast to the electrons which are extreme relativistic, and we also took into account that before the reaction $p + e \rightleftarrows n$ took place, $N_e/N_\nu = 0.5$. The above relations allowed us to also determine the relative contributions to the total kinetic energy of neutrons, protons, and electrons after the equilibrium of the reaction[3] $p + e \rightleftarrows n$ has been established: $E^{'}_{\mathrm{Kn}} : E^{'}_{\mathrm{Kp}} : E^{'}_{\mathrm{Ke}} = 0.9314 : 0.0011 : 0.0675$. Equating $E^{'}_{\mathrm{Kn}} = b\hbar^2(N_\nu x)^{5/3}/m_n R^2$ to $0.9314\frac{1}{2}E_G = 0.9314\frac{1}{2}aGm_u^2N_\nu^2/R$ we obtain the radius of the neutron star,

$$R = 1.92(b/a)(m_e/m_n)(e^2/Gm_u^2)N_\nu^{-1/3}a_B \qquad (12.3)$$

which for $b/a = 4.51/2$ and $N_\nu = 1.43\ N_{\nu,S}$ gives $R = 13$ km.

As in the white dwarf case, the kinetic energy per nucleon in neutron stars is proportional to $N_\nu^{4/3}$. Thus for very large mass even the neutrons and the protons will reach the extreme relativistic limit, which means that the total kinetic energy will be of the form B/R with $B = b_1\hbar c N_\nu^{4/3}(x^{4/3} + 2y^{4/3})$, where $N_n' = xN_\nu$ and $N_p' = N_e' = yN_\nu$, with x and y being different than before. The reason is that in this extreme relativistic limit the equilibrium equation for the reaction $p + e \rightleftarrows n$, $\mu_n = \mu_p + \mu_e$, becomes $N_n'^{1/3} = N_p'^{1/3} + N_e'^{1/3} = 2N_p'^{1/3}$ which implies that the ratios of neutrons to protons and to electrons will be as follows: $N_n' : N_p' : N_e' = 8 : 1 : 1$; since $N_n' + N_p' = N_\nu$, it is found that $x = 0.8888$ and $y = 0.1111$. Then the coefficient B becomes $B = 0.9615 b_1 \hbar c N_\nu^{4/3}$. Equating this with $aGN_\nu 2m_u^2$ we obtain

$$N_{\nu_2} = [0.9615(b_1/a)(\hbar c/Gm_u^2)]^{3/2} \qquad (12.4)$$

which for $b_1/a = 2.126$ gives $N_{\nu 2} = 5.5N_{\nu,S}$. The actual critical value seems to be about half this result.

[4] The neutrinos and antineutrinos produced in these reactions will not be retained within the system.

12.4 Black Holes

As we did in the case of white dwarfs and neutron stars, we shall restrict ourselves to the simpler case of an electrically neutral, *non-rotating* black hole. Such a black hole is accompanied by a spherical surface, called the *event horizon*. Physically, the event horizon is the perfect one-way surface in the sense that, while it permits all kind of particles to cross it from its exterior, no particle, not even photons, is allowed to escape from its interior. Nevertheless, it is believed that a black hole can emit radiation, in particular E/M radiation, from a region outside but near the event horizon by a quantum process proposed by Hawking. The process is based on the quantum nature of vacuum according to which particle/antiparticle pairs appear and disappear all the time existing for a very short time $\Delta t \approx \hbar/\Delta\varepsilon$, where $\Delta\varepsilon$ is the total energy of the pair. Photon pairs having zero rest mass are the easiest to appear/disappear. In the region just outside the horizon of the black hole the gravitational field is so strong that one partner of the quantum fluctuation pair may be sucked in by the black hole leaving the other partner unpaired and, hence, transformed from a virtual to a real particle. The energy during this process is conserved, since the energy of the created real particle comes at the expense of the total energy of the black hole. A detailed calculation of this process shows that the frequency distribution of the emitted E/M radiation by the black hole is that of a black body with an effective temperature T. Thus a temperature and, therefore, an entropy is attributed to a black hole. A non-rotating, electrically neutral black hole is characterized only by its mass M. All other related quantities such as, its total energy $U = Mc^2$, its event horizon radius R_S, called also the Schwarzchild radius, its temperature T, its entropy S, and its life-time t, are all functions of its mass. Of course, we expect that these functions will contain the following universal physical constants: The gravitational constant G (since black holes are par excellence gravitationally controlled entities), the velocity of light c (since black holes are governed by extreme relativistic laws), and Planck's constant $\hbar$ (where quantum mechanical effects play a role). Having these comments in mind and employing dimensional considerations, we obtain the following result for the radius of the event horizon:

$$R_S = \eta GM/c^2 \tag{12.5}$$

General relativity shows that $\eta = 2$; accidentally, the classical calculation of the escape velocity from the event horizon of a particle having velocity c gives also $\eta = 2$. The temperature of a black hole, appearing in the combination $k_B T$, must depend on the gravitational field strength and, hence, on the product GM, on $\hbar$ (as a quantum phenomenon) and on c (as a relativistic phenomenon). The only combination of the last three quantities having dimensions of energy is $\hbar c^3/GM = 2\hbar c/R_S$. Thus

$$k_B T = \eta' \hbar c/R_S \tag{12.6}$$

It turns out that $\eta' = 1/4\pi$, where the factor 1/4 is coming by averaging cos θ over half the solid angle. The cos θ factor appears, because the emitted radiation is proportional to the outgoing normal component of the flux. The entropy of a black hole can be obtained by the first law of thermodynamics, $TdS = dU = c^2 dM$, (12.6), and (12.5). We find

$$S = 4\pi k_B GM^2/\hbar c = \pi k_B c^3 R_s^2/\hbar G \tag{12.7}$$

Notice that the entropy is proportional to the area of the event horizon and not to the volume enclosed by it, as in ordinary thermodynamic systems (for comments on the entropy of black holes see 12.2). Finally, the life-time of a black hole of initial mass M_o can be obtained from the equation $c^2 dM/dt = -4\pi R_S^2 \sigma T^4$, where σ, the Stefan–Boltzmann constant, has been defined in Sect. 11.2. Substituting from (12.5) and (12.6) we find that

$$M^3(t) = M_o^3 - a[\hbar c^4/G^2]t, \quad a \equiv \pi^2/10(8\pi)^3 \tag{12.8}$$

from which the life-time follows immediately by setting $M(t) = 0$.

Notice that the three universal constants $\hbar$, c, and G define a system of units, the so-called Planck's system with length unit $\ell_P = \sqrt{G\hbar/c^3} = 1.6161 \times 10^{-35}$ m, time unit $t_P = \sqrt{G\hbar/c^5} = 5.3907 \times 10^{-44}$ s and mass unit given by $m_P = \sqrt{\hbar c/G} = 2.1766 \times 10^{-8}$ kg. The energy unit is $\varepsilon_P = m_P c^2 = 1.2234 \times 10^{28}$ eV and the temperature unit is $T_P = 1.4196 \times 10^{32}$. In Planck's units formulae (12.5)–(12.8) become simpler and easier to use. In particular, the entropy becomes $S/k_B = \frac{1}{4}\tilde{A}$, $\tilde{A} = 4\pi\tilde{R}_S^2$, $\tilde{R}_S = R_S/\ell_P$.

12.5 The Minimum Mass of an Active Star

In Fig. 12.1 we plot the total and the kinetic energy as well as the temperature as a function of the radius R of a gaseous body in its way to become a star. Whether this will happen depends on the maximum temperature T_M being larger than the ignition temperature T_i for the $p + p \rightarrow pn + e^+$ fusion reaction to start taking

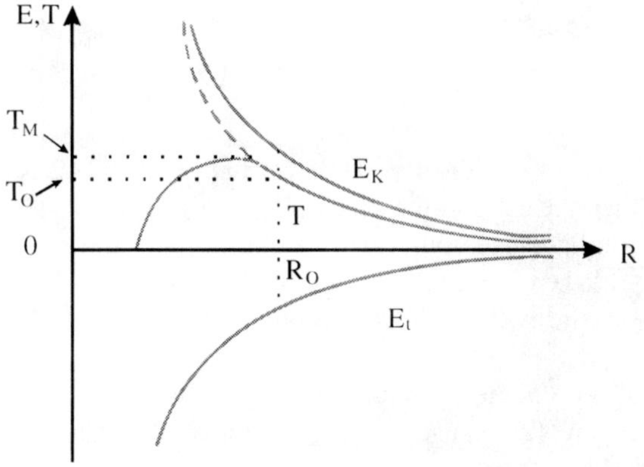

Fig. 12.1 The total energy E_t, the kinetic energy E_K, and the temperature T (*solid line*), of a star to be is shown as a function of its radius R. The *dash line* gives the classical result for the temperature which coincides with the quantum one for large values of R. T_o is the equilibrium average value of the temperature and R_o that of the radius

place. Notice that the existence of the maximum temperature T_M stems from the quantum kinetic energy. Classically, the temperature will be proportional to the kinetic energy according to the formula $E_{Kc} = \frac{3}{2}Nk_BT$ and will exhibit no maximum, while for small R and low T quantum mechanics gives $E_{Kq} = b\hbar^2 N_e^{5/3}/m_e R^2$. Interpolating between these two limits we obtain a simple approximate formula for the exact quantum expression $E_K = f(R, T)$:

$$E_K = \left(E_{\text{Kc}}^2 + E_{\text{Kq}}^2\right)^{1/2} \tag{12.9}$$

The virial theorem implies that $E_K = -\frac{1}{2}E_G$ where $-E_G = aGM^2/R = aGm_u^2 N_v^2/R$. Assuming full ionization of all the atoms in the star to be, we have $N = N_e + N_a = xN_v$ and $N_e = yN_v$, where a reasonable estimate of the composition gives $x \approx 1.5$ and $y \approx 0.8$. By combining (12.9) and $E_K = -\frac{1}{2}E_G$ we obtain k_BT as a function of R. By differentiating this function with respect to R we find an approximate expression for k_BT_M:

$$k_BT_M \approx \gamma(Gm_u^2/e^2)^2(e^2/a_B)N_v^{4/3}, \quad \gamma = a^2/12bxy^{5/3} \tag{12.10}$$

By choosing $b/a \approx 2$, $a \approx 0.8$, (see (12.1)) we find $\gamma \approx 0.038$. The ignition energy k_BT_i can be estimated by dimensional considerations: it must depend on e^2, since it is the Coulomb repulsion which demands high kinetic energy for two protons to approach each other; it must also depend on the reduced mass of the two protons $m_p/2$ and Planck's constant $\hbar$, since the fusion is greatly facilitated by the quantum tunneling effect. Hence,

$$k_BT_i = \eta e^4 m_p/2\hbar^2 \tag{12.11}$$

where the numerical constant η is expected to be of the order of 0.1 or less, since the Boltzmann distribution allows velocity squares much larger than the average value. Actually, since the temperature 1.57×10^7 K at the center of the sun allows fusion, it means that the ignition temperature must be clearly less than that, which implies that $\eta < 0.054$. Equating (12.10–12.11) we obtain for the minimum mass of a star

$$M_1 = N_1 m_u \approx \left(\frac{\eta}{2\gamma}\right)^{3/4}\left(\frac{m_p}{m_e}\right)^{3/4}\left(\frac{e^2}{Gm_u^2}\right)^{3/2} m_u \tag{12.12}$$

By choosing $\eta = 0.04$, we find $M_1 \approx 2.42 \times 10^{56} m_u$ i.e. five times smaller than the mass of the sun, while stars as small as ten times less than the mass of the sun have been observed.

When the mass M of the star is larger than M_1, the temperature at the center of the star will reach the ignition temperature for $R \approx R_o$, where R_o is the equilibrium value of R, as shown in Fig. 12.1. Therefore, the average equilibrium value T_o of the temperature is proportional to the ignition temperature: $T_o = zT_i$. Figure 12.1 suggests that for $R \approx R_o$ the classical result $\frac{3}{2}xN_v k_BT_o$ is a reasonable approximation to the kinetic energy E_K; the latter is equal to $-\frac{1}{2}E_G = \frac{1}{2}aGm_u^2N_v^2/R_o$. Hence, we end up with the equality $\frac{1}{2}aGm_u^2N_v^2/R_o \approx \frac{3}{2}xN_vk_BzT_i$ from which the proportionality $R_o \propto N_v$ follows. More explicitly we have taking into account (12.11) too:

$$R_o \approx \frac{2a}{3xz\eta}\frac{Gm_u^2}{e^2}\frac{m_e}{m_p}N_v a_B \tag{12.13}$$

The proportionality factor z can be determined by comparing with the solar data: we find $z \approx 0.35$, a reasonable value, since T_o is the average temperature over the whole star and not its value at the center of the star where the highest temperatures are expected. Using the previously estimated values of a, x, y, and η, we obtain the following formula for the radius of an active star roughly the size of the sun: $R_o \approx 5.837 \times 10^{-49}N_v$ m. In reality, for stars more massive than $2M_S$ but less massive than $20M_S$ the kinetic energy increases with the mass a bit faster than linearly, which in turn, implies that the radius/mass relation, becomes sublinear: $R_o \propto N_v^s, s \approx 0.75$ or even less. In the range $0.5M_S < M < 20M_S$ the surface temperature T is roughly proportional to $M^{1/2}$ so that the luminosity L ($L = 4\pi R_o^2 I$; I is defined in p. 35.) which is proportional to $R_o^2T^4$ becomes proportional to $M^{2s}M^2 = M^{2(1+s)}$. We have seen already that $s \approx 1$ for about 0, $5M_S < M < 2M_S$; thus for this range of masses $L \propto M^4$. For M larger than about 2 but less than about 20 solar masses $s \approx 0.75$ and, consequently, $L \propto M^{3.5}$. The luminosity determines the life-time t of an active star. Indeed, assuming that the available energy to be emitted is proportional to the nuclear fuel which in turn is proportional to the initial mass of the star, we can conclude that $t \propto M/\bar{L} \propto M^{-1-2s}$.

12.6 The Maximum Mass of an Active Star

There are very few observational data regarding the upper mass limit of an active star for two reasons: First, such big stars need special conditions for their creation. Second, even if they manage to be formed, they die young, since $t \propto M^{-2.5}$. Thus very few of them are present at any given time. From the theoretical side, the following argument is usually advanced: High mass implies high temperature, which in turn produces high radiation pressure, P_R, due to its T^4 dependence versus the linear on T dependence of the thermal kinetic pressure. Thus, for high enough mass the radiation pressure P_R will exceed the thermal kinetic pressure P_K. When $P_R = \Lambda P_K$, where Λ is usually taken around 2, an instability is expected to occur and, as a consequence, the star will cease to exist. Replacing in the equality $P_R = \Lambda P_K$, the expressions for $P_K = nk_BT = xn_vk_BT$ and $P_R = \frac{1}{3}U_R/V$, (see (5.4), p. 35), where $n_v \equiv N_v/V$ we obtain

$$(k_BT)^3 = (45\Lambda/\pi^2)(\hbar c)^3 xn_v \tag{12.14}$$

The gravitational pressure P_G is determined by differentiating $E_G \propto V^{-1/3}$ with respect to $-V$ and it is found to be $E_G/3V = -(4\pi/3)^{1/3}(a/3)Gm_u^2 n_v^{4/3} N_v^{2/3}$. Substituting the above expressions of the pressures in the equilibrium condition $|P_G| = P_K + P_R = P_K(1 + \Lambda)$ we have:

$$k_B T = (4\pi/3)^{1/3}(a/3)Gm_u^2 n_v^{1/3} N_v^{2/3}/(1+\Lambda)x \qquad (12.15)$$

Eliminating $k_B T$ between (12.14) and (12.15) we obtain an equation for the maximum value of N_v to be denoted as N_2:

$$N_2 = \left(\frac{135\Lambda}{4\pi^3}\right)^{1/2}\left(\frac{3(1+\Lambda)}{a}\right)^{3/2} x^2 \left(\frac{\hbar c}{Gm_u^2}\right)^{3/2} \qquad (12.16)$$

For $\Lambda = 2$ and the other parameters as before we find $N_2 = 2.82 \times 10^{59}$ or $N_2/N_S = 235$, while for $\Lambda = 1$ we find $N_2 = 1.08 \times 10^{59}$ or $N_2/N_S = 91$.

Chapter 13
The Observable Universe

Abstract Hubble's recession of the distant galaxies, the observation of the cosmic microwave radiation, and recent observational data coupled with Einstein's gravitational theory have established the inflationary Big Bang model as the standard scheme for determining the important milestones in the evolution of the Universe.

Keywords Expansion · Big Bang · Inflation · CBR · Light and matter dominated epoqs

Summary and Outline The theoretical study of the Universe is based on the assumption that it is homogeneous, isotropic, and without boundaries. The assumed homogeneous and isotropic nature of the Universe is considered justified, if we ignore "details" of the order of 500 million light years. In other words, we restrict ourselves to a coarse description of the Universe by averaging over distances much larger than 100 million light years. Both observations and theory show that the Universe is expanding—space itself is expanding—at a rate $\dot{R} \equiv dR/dt$ which is proportional to the distance R between two points.[1] The proportionality coefficient is called Hubble's constant and is denoted by the letter $H \equiv \dot{R}/R$. The basic equation determining the time dependence of H is the following:

$$\left(\dot{R}/R\right)^2 = \left(8\pi G/3c^2\right)\varepsilon \tag{13.1}$$

where ε is the average energy density in the Universe. Notice that both H and ε are independent of R (because of the homogeneity assumption), but they do depend on time. This picture shows that in the past the Universe was denser and hotter until

[1] Clusters of galaxies containing hundreds of galaxies and smaller structures than that do not follow space in its expansion, because gravitational attraction keeps them together.

E. N. Economou, *A Short Journey from Quarks to the Universe*,
SpringerBriefs in Physics, 1, DOI: 10.1007/978-3-642-20089-2_13,

we would reach a time zero where the whole Universe was almost infinitely small, dense, and hot. From this zero point on, the Universe is expanding and cooling, a process termed *Big Bang*. By applying the known laws of physics, we can follow the history of the Universe (in coarse terms) and reach conclusions regarding its 'present' features, such as the value and the composition of the energy density, the value of Hubble's constant, the percentage of chemical elements, the radius of the observable Universe, the spectral and angular distribution of cosmic photons, larger structures of the Universe such as galaxy cluster, voids, filaments etc., etc. Notice that time 'present' may include time 'past' as well, since by observing now an event at a distance d from us, we actually get what was happening there at a time $t \approx d/c$ before now. Comparison of the theoretical and numerical results with a stream of very impressive observational data has enabled the human civilization for the first time to provide some reliable answers to the age old, yet ever present question: How did the World start and how did it evolve?

In this chapter we start with a few historical remarks regarding the field of Cosmology and the recent impressive observational data. Then, we "derive" the basic equation (13.1) by employing Newtonian mechanics, the results of which can be trusted as long as they agree with those of the general theory of relativity. By taking into account some observational data regarding the composition of the energy density of the Universe, we analytically solve (13.1) (making some approximations in order to avoid numerical solutions). We determine this way the size and the temperature of the Universe as a function of time elapsed since the time zero of the Big Bang. Having as input the temperature and the density as well as the known laws of Physics we reconstruct in coarse terms some of the basic events in the history of the Universe.

13.1 Introductory Remarks

The modern examination of the Universe really started in 1917 with Einstein finding a time independent, but unstable solution of the metric of the Universe within the framework of his general theory of relativity (GTR) in which he included the so-called *cosmological constant*. Alexander Friedmann, Georges Lemaitre and others found during the 1920s time dependent solutions of the metric of the Universe exhibiting Big Bang characteristics with the expansion either continuing for ever or terminating and reversing direction leading eventually to a Big Crunch. This type of solutions received a substantial boost by an extremely important observational fact, namely Hubble's discovery that distant galaxies go away from Earth with a velocity proportional to their distance from Earth. This discovery strongly indicated that the Universe is expanding as predicted by the time dependent solutions of the general relativity equations.

The next big observational discovery followed the theoretical work of Gamow et al. in 1948 anticipating the existence of a relic electromagnetic radiation extending almost uniformly and isotropically over the whole Universe. This so-called *cosmic background radiation* (CBR) was serendipitously discovered by Penzias and Wilson in 1964 and played a crucial role in transforming the field of Cosmology from what most physicists considered as kind of metaphysics to a genuine physical subject. Following the initial discovery of the CBR, a much more extensive and precise spectral analysis of the CBR was performed by a team headed by Smoot and Mather through the instrument FIRAS mounted on the COBE satellite. The analysis has shown that the frequency distribution of the CBR was in excellent agreement with that of a black body of temperature 2.725 K. Moreover, the COBE mission detected a minute anisotropy in the temperature of 2.725 K (of the order of one part in a 100,000), indicative of very small inhomogeneities in the early Universe.

Inspired by the COBE success many experiments were designed aiming at measuring this minute anisotropy as accurately as possible. Probably the most impressive among them is the one obtained through the WMAP mission which measured the CBR temperature fluctuations as a function of the direction of observation, i.e., as a function of the angles θ, and $\phi{:}\delta T(\theta, \phi)$ This function was written as a weighted sum of the functions $Y_{\ell m}(\theta, \phi)$ (we met these functions in Chap. 8, when we examined the angular dependence of the atomic orbitals). The average of the square of the absolute value of the coefficients of this expansion, denoted as C_l, is plotted in Fig. 13.1.

The main conclusions resulting by the combination of theory and observational data are the following:

1. The expansion of the Universe was decelerated in the past and is accelerated presently. This implies that a cosmological constant-like term enters in the equations of general relativity or, equivalently, the energy density contains a contribution almost time-independent in spite of expansion of the Universe.
2. The energy density of the Universe consists of several contributions:

$$\varepsilon = \varepsilon_b + \varepsilon_e + \varepsilon_\nu + \varepsilon_{\rm dm} + \varepsilon_{\rm ph} + \varepsilon_{\rm de}. \tag{13.2}$$

The contribution ε_b is due to ordinary matter (protons, and neutrons bound, presently, in nuclei); ε_e is due to electrons and it is presently negligible in comparison to ε_b; ε_ν is coming from neutrinos and antineutrinos, otherwise called light dark matter; $\varepsilon_{\rm dm}$ is associated with the so-called dark matter, or more precisely heavy dark matter (several exotic particles predicted by some proposed theories, but not yet found experimentally, are candidates for making up the dark matter); $\varepsilon_{\rm ph}$ is due to the photons; finally, $\varepsilon_{\rm de}$ is the so-called dark energy contribution which seems to be almost or exactly time-independent (in what follows we shall consider $\varepsilon_{\rm de}$ as exactly time-independent, which means that it plays a role equivalent to the cosmological constant). The nature of $\varepsilon_{\rm de}$ is a mystery.

A clarification is needed regarding the number of electrons, protons, neutrons, neutrinos, etc. When the temperature of the Universe was much larger than $T = m_e c^2/k_B = 5.93 \times 10^9$ K, the reaction $e + e^+ \rightleftarrows n\gamma$ was possible, where e^+ is

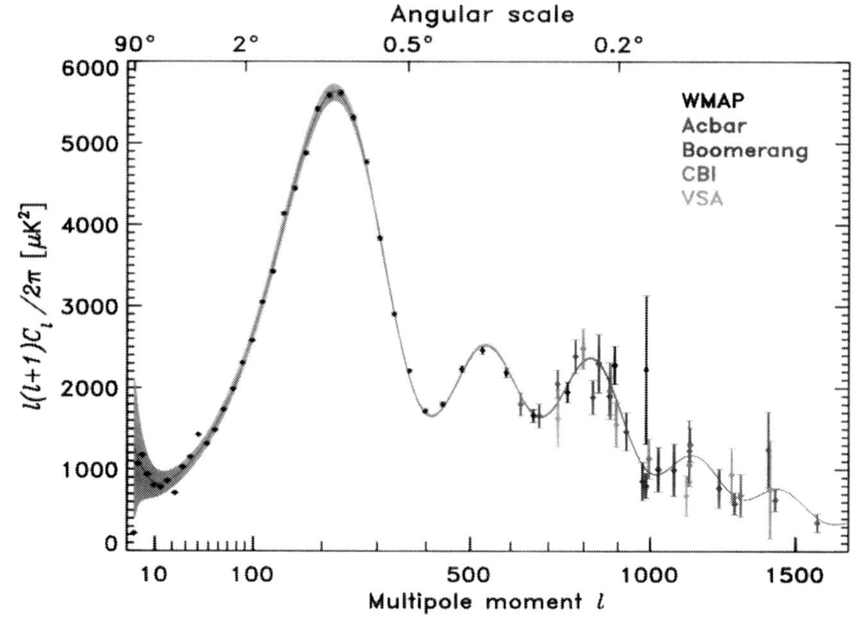

Fig. 13.1 The quantity C_l, defined above, multiplied by $l(l+1)/2\pi$ is plotted vs. the index l (NASA's webpage). The measurements are denoted by points and error bars; different colors indicate measurements by different instruments and techniques. The continuous line shows the best fit to the data according to the prevailing theory which contains a minimum of seven adjustable parameters, such as the various components of the total energy density, the Hubble constant, etc. The oscillatory character of the above line implies damped acoustic oscillations in the early Universe driven by the gravitational attraction and the photonic repulsion with the peaks representing the fundamental and the higher harmonics. These acoustic oscillations were evolved from quantum fluctuations in the very early Universe. The position of the first peak implies that the geometry of the Universe is Euclidean. The second and the third peaks determine the ratio of the baryonic mass density to the so-called dark mass density

the positron and n is the number of photons involved in the reaction (usually $n = 2$). As it was mentioned in Sect. 12.3, thermodynamic equilibrium implies a relation among the chemical potentials of the species involved in a reaction. For the above reaction the relation is $\mu_e + \mu_{e^+} = n\mu_\gamma = 0$. This last equation together with the equation $n_e - n_{e^+} = n_\ell$ (expressing the fact that the difference in the concentrations of electrons and positrons is equal to the concentration of electron lepton number n_ℓ) allows the calculation of n_e, n_{e^+}, ε_e, and ε_{e^+} as a function of the temperature and the electron lepton number concentration n_ℓ. Notice that both the lepton number and the baryon number $N_b = V\ n_b$ are conserved. Moreover, since eventually the neutron to proton ratio is 1:7 as shown in Sect. 13.4, electrical neutrality implies that $n_\ell = 0.875 n_b$. We shall show below that the photon number is also approximately conserved and that the ratio $N_b/N_{\rm ph} \approx 0.6 \times 10^{-9}$ remains almost constant

Table 13.1 Present ($t \simeq 13.7$ billion years) composition of the Universe

Kind of matter	Kind of particles	Particle mass (eV)	Number of particles in the observable Universe	Percentage in the energy of the Universe %	Percentage in pressure %	Percentage in energy at $t =$ 379, 000 years (%)
Baryonic	Proton Neutrons (in nuclei)	$\approx 10^9$	0.86×10^{80}	$4.6\% \pm 0.1\%$	$\simeq 0$	12
Light dark matter	Neutrinos	$\ll 1$	10^{89}	$< 1\%$	$\simeq 0$	10
Heavy dark matter	Supersymmetric particles ?	10^{11}?	5×10^{78}?	$23.3\% \pm 1.3\%$	$\simeq 0$	63
E/M radiation	Photons	$6.34 \cdot 10^{-4}$	1.4×10^{89}	0.005%	$\simeq 0$	15
Dark energy	? (Scalar particles?)	? (10^{-32}?)	? (10^{122}?)	$72.1\% \pm 1.5\%$	100%	≈ 0

Radius of the observable Universe: 46 billion light years = 4.35×10^{26} m

throughout the history of the Universe. Such a small number of this ratio makes the following proposal very attractive: Initially, N_b, and N_ℓ were exactly zero, so that matter and antimatter was created out of electromagnetic energy through the reaction photons $\rightleftarrows$ matter + antimatter; then a mechanism breaking to a minute degree the symmetry between matter and antimatter is assumed to intervene so that one part in a billion of matter is left out after the annihilation. Before the annihilation, i.e., for $k_BT \gg m_ec^2$, the electron lepton number is negligible in comparison with N_e or N_{e^+}; hence, $n_e \approx n_{e^+}$ and the result for the energy density is $\varepsilon_e \approx \varepsilon_{e^+} \approx \frac{7}{8}\varepsilon_{\text{ph}}$ (see problem 13.2). In the opposite limit, $k_BT \ll m_ec^2$, all the positrons have been annihilated and the left out electrons are equal to the electron lepton number. A similar to electron/positron process, but much more complicated, has taken place with the quark/antiquark and eventually with the proton/antiproton, neutron/antineutron pairs with the threshold temperature being, of course, associated with the rest energies of the proton, and neutron, $T = m_pc^2/k_B = 1.09 \times 10^{13}$ K. The neutrino/antineutrino pair is similar to the electron/positron pair but with two important differences. The neutrino mass is extremely small so that the corresponding threshold temperature may be 1,000 K or even lower. The neutrino/antineutrino pair is interacting with everything else through the weak interaction, so that neutrinos and antineutrinos became decoupled from the rest at an early stage ($T \approx 10^{10}$ K) and their number remains constant and comparable to the photon number since then.

Presently, the various contributions to the total energy density of the Universe, ε, are as follows: $\varepsilon_b : \varepsilon_\nu : \varepsilon_{\text{dm}} : \varepsilon_{\text{ph}} : \varepsilon_{\text{de}} \approx 4.6 : 0.4 : 23 : 0.005 : 72$ (See Table 13.1 and Table 13.2 for the values of other relevant quantities). The present neutrino contribution is quite uncertain, since the neutrino mass is not known. The neutrino temperature is calculated to be 71.4% of the photon temperature [13.1].

Table 13.2 "Best" values of some cosmological parameters

Total density	$\Omega_{\text{tot}} \equiv \varepsilon/\varepsilon_{cr} = 1.02 \pm 0.01$
Critical density	$\rho_{cr} \equiv \varepsilon_{cr}/c^2 = 3H_o^2/8\pi G = 1.034 \times 10^{-26}\,\text{kg/m}^3 \simeq 6.18\,m_p/\text{m}^3$
Baryon density	$\rho_B \equiv \varepsilon_B/c^2 = 4.7 \times 10^{-28}\,\text{kg/m}^3 = 0.28\,m_p/\text{m}^3$
Hubble constant	$H_o \simeq 0.24 \times 10^{-17}\text{s}^{-1} = (0.742 \pm 0.036)\,100\,\text{km/s} \times \text{Mpc}$
Age of the Universe	$t_o = 13.73 \pm 0.12\,\text{Gyr}$
Age of decoupling	$t_{\text{dec}} = 379\,\text{kyr}$
Age of first star	$t_s \simeq 108\,\text{Myr}$
Radius of observable Universe	$R_o \simeq 4.35 \times 10^{26}\,\text{m}$
Expansion since decoupling age	$R_o/R_{\text{dec}} = 1089$
Volume of observable Universe	$V_o \simeq 3.45 \times 10^{80}\,\text{m}^3$
Ratio of photons to baryons	$N_{ph}/N_B \approx 1.64 \times 10^9$
Temperature of CBR	$T_{\text{CBR}} \approx 2.725\,\text{K}$

13.2 Derivation of the Equations Determining the Expansion Rate

We shall try first the dimensional analysis for Hubble's constant, whose dimension is one over time; we shall also follow the same approach for the relative acceleration $\ddot{R}/R$. Obviously, the gravitational constant G must enter in the formulas for $\dot{R}/R$ and $\ddot{R}/R$; we expect that the total mass density $\rho = \varepsilon/c^2$ must also enter, since the gravitational field involves the product G times mass and hence the product $G\rho$. However, since the dimension of $G\rho$ is one over time square, we can conclude that the most general relation for $\dot{R}/R$ and $\ddot{R}/R$ are of the form

$$\dot{R}/R = \sqrt{G\rho}f_1(x_1, x_2, \ldots) \quad \text{and } \ddot{R}/R = G\rho f_2(x_1, x_2, \ldots) \tag{13.3}$$

where $x_1, x_2, \ldots$ are dimensionless variables. One obvious such variable is the ratio of the pressure over the energy density, $x_1 = p/\varepsilon$; another is the ratio $x_2 = (c^2/R^2)/G\rho$ of c^2/R^2 over $G\rho$. This last variable x_2 will not appear in $\dot{R}/R$, if the latter is really constant, i.e., independent of R. Let us assume that there are no other dimensionless variables and let us proceed with the determination of f_1 and f_2.

Although the reliable way to obtain these functions is to employ the GTR, here we shall work within the Newtonian framework. The latter at closer scrutiny reveals inconsistencies in some of the formulae obtained; nevertheless, the main results are in agreement with the conclusions of the GTR. This allows us to employ the Newtonian approach, taking advantage of its simplicity and familiarity, and accept only its results which agree with those of the GTR.

Consider a far away galaxy of mass m at a distance R from the observation point. Newton's equation, $m\ddot{R} = F$ becomes $\ddot{R} = -GM/R^2$, where M is the mass enclosed within the sphere of radius R centered at the observational point. By integrating this equation, *assuming constant mass M*, we obtain a relation showing that the sum of kinetic and potential energy is a constant, $\frac{1}{2}C\,m$

$$\frac{1}{2}\dot{R}^2 - \frac{GM}{R} = \frac{1}{2}C \tag{13.4}$$

Equation (13.4) can be brought to a form similar to (13.1) by the following operations: Substitute $\frac{4\pi}{3}R^3\rho$ for M; divide by $R^2/2$; replace ρ by ε/c^2; and write $\varepsilon = \varepsilon' + \varepsilon_{\mathrm{de}}$. We have then

$$\left(\dot{R}/R\right)^2 = \left(8\pi G/3c^2\right)\varepsilon - \left(C/R^2\right) = \left(8\pi G/3c^2\right)\varepsilon' + \left(8\pi G/3c^2\right)\varepsilon_{\mathrm{de}} - \left(C/R^2\right) \tag{13.5}$$

The separation of the total energy density ε in two terms ε' and $\varepsilon_{\mathrm{de}}$ makes sense, if the latter, as opposed to the former, is a universal constant. In this case we can define $(8\pi G/3c^2)\varepsilon_{\mathrm{de}} \equiv \Lambda/3$, where Λ is the so-called cosmological constant. Equation (13.5), with the $\varepsilon_{\mathrm{de}}$ term assumed constant and written in terms of the cosmological constant, is exactly the outcome of Einstein's equations of GTR in combination with the assumptions of homogeneity and isotropy. In what follows we shall write (13.5) by setting $(8\pi G/3c^2)\varepsilon' + (8\pi G/3c^2)\varepsilon_{\mathrm{de}} = (8\pi G/3c^2)\varepsilon$. It turns out that the constant C appearing in (13.5) determines the type of geometry of space: If C is positive, the geometry of the three-dimensional space is of positive curvature, the analogue of the two-dimensional geometry on the surface of a sphere; if C is negative, the geometry is of negative curvature, the analogue of the two-dimensional geometry on a saddle-like surface. Finally, if $C = 0$, the geometry is of zero curvature, i.e., it is the familiar Euclidean geometry; in this case we say that the three-dimensional space is flat. Notice that for $C = 0$, (13.5) reduces to (13.1). The $C = 0$ choice, which implies that the total energy as shown in (13.4) is zero, is supported by the *inflation* model. This model assures that just after time zero, more precisely during the period $t \approx 10^{-35}$ to $t \approx 10^{-33}$s, there was an exponentially fast expansion, a true 'Big Bang', which increased the distance between any two points by a factor of at least 10^{30}! There are proposed theories, of quantum nature, involving phase transitions among elementary particles, to account for this dramatic expansion process. But the main reason that inflation is tentatively accepted and incorporated in the "standard" cosmological model, is that it explains many puzzling data, such as the same temperature of the CBR coming from points of opposite directions relative to Earth and near the edges of the observable Universe,[2] or the Euclidean geometry[3] as deduced by the data of Fig. 13.1, or the appropriate minute size of the quantum fluctuations during the inflation period acting as seeds for the eventual formation of galaxies, protostars, etc. Notice, however, that there is a minority of experts who attempt to explain these puzzling data by other approaches than those of the inflation model. Anyway, one must keep in mind that events at time $t \approx 10^{-35}$s after the Big Bang involve energies as high as 10^{23} eV per particle, i.e., ten orders of magnitude higher than the upper limit achieved in the biggest accelerators on Earth; thus a direct experimental confirmation of such events is missing. Furthermore, a reliable theoretical study of events where both quantum and gravitational effects are involved requires a complete quantum theory of gravity, which is not yet available.

After the end of the inflation period it is expected that the evolution of the Universe obeyed GTR, i.e., (13.1). From this equation and the first law of thermodynamics, $dU = -pdV$, one[4] can obtain an equation for the quantity $\ddot{R}/R$ which is in agreement with the corresponding one of the GTR. The proof follows: Taking into account that $U = V\varepsilon = \frac{4\pi}{3}R^3\varepsilon$, the first Law shows that

$$d\varepsilon/dR = -(3/R)(\varepsilon + p) \quad (13.6)$$

If we multiply (13.1) by R^2 and then differentiate it with respect to time we find $2\dot{R}\ddot{R} = (8\pi G/3c^2)\left[2R\dot{R}\varepsilon + R^2\dot{\varepsilon}\right]$. If in this last equation we replace the quantity $\dot{\varepsilon} \equiv d\varepsilon/dt$ by $(d\varepsilon/dR)\dot{R} = -(3/R)(\varepsilon + p)\dot{R}$, where (13.6) was taken into account, we obtain

$$(\ddot{R}/R) = -(4\pi G/3c^2)(\varepsilon + 3p) \quad (13.7)$$

Notice that (13.7) is not consistent with our initial Newtonian equation $\ddot{R} = -GM/R^2$, unless $p = 0$: This shows that results based on the Newtonian approach are to be trusted only if they agree with the ones based on the GTR. Another point to be noticed is that the constancy of any component of ε, such as the dark energy component, implies, according to (13.6), that $p_{\mathrm{de}} = -\varepsilon_{\mathrm{de}}$. Taking out of ε and p the corresponding dark energy components and recalling the relation $(8\pi G/3c^2)\varepsilon_{\mathrm{de}} \equiv \Lambda/3$, we can rewrite (13.7) as follows:

---

[2] If the rate of expansion predicted by the GTR was the actual one at *all times*, the distance between these points would be at all times t much larger than $c\,t$; hence these points would never have the opportunity to communicate with each other and established the same temperature; thus the coincidence of their temperatures would be an accident. In contrast, the existence of such a fast expansion rate as that of inflation, implies that these points were in close communication before the inflation period and, hence, they had the opportunity to have equal temperature.

[3] An expansion by thirty orders of magnitude in almost zero time will make the term C/R^2 60 orders of magnitude smaller and, hence, absolutely negligible.

[4] The term TdS is practically zero, because the dissipation terms are negligible in comparison with the energy density which contains the rest energy of the particles.

$$\left(\ddot{R}/R\right) = -\left(4\pi G/3c^2\right)\left(\varepsilon' + 3p\right) + \tfrac{1}{3}\Lambda \tag{13.7a}$$

13.3 Solution of the Equation for the Expansion Rate

To solve either (13.1) or (13.7) we need the dependence on R of each component of ε or p, where from now on by R we shall denote the radius of the observable Universe.

For $\varepsilon_{\mathrm{de}}$, following the observational evidence, we shall assume that it does not depend on R and, hence, that it is equal to $-p_{\mathrm{de}}$.

For $\varepsilon_{\mathrm{ph}}$ and p_{ph} all the information needed have been obtained in Sect 5.1: $p_{\mathrm{ph}} = \frac{1}{3}\varepsilon_{\mathrm{ph}} = \frac{\pi^2}{45}(k_BT/\hbar c)^3 k_BT$ and $S_{\mathrm{ph}} = \frac{4}{3}V\varepsilon_{\mathrm{ph}}/T = \frac{16\pi^3}{135}k_B(Rk_BT/\hbar c)^3$; the number of photons can be obtained by dividing the total energy $V\varepsilon_{\mathrm{ph}}$ by the average energy of each photon which is ak_BT, where a turns out to be $a \approx 2.7$, $N_{\mathrm{ph}} \approx 1.02(Rk_BT/\hbar c)^3$. Since the entropy remains almost constant, it follows that the number of photons is also almost constant and that $T_{\mathrm{ph}} \propto 1/R$ and $\varepsilon_{\mathrm{ph}} \propto 1/R^4$.

The R dependence of the hadronic contributions to ε and p is complicated because reactions due to strong interactions are taking place for temperature higher than about 10^{13} K. Below this temperature the quark/gluon plasma has been transformed to baryons, the antibaryons have been annihilated, and the baryonic contributions to the energy density, $\varepsilon_b \approx N_b m_B c^2/V \propto 1/R^3$, and the pressure, $p_b \approx \frac{3}{2}N_b k_B T$, are negligible. Baryons will become important much later, at $t \geq 300$ kyr, when the temperature is about 3000 K.

For the joint electron and positron contribution we have

$$\begin{aligned} &\text{If } a'k_BT \gg m_ec^2, \quad \text{then } 2p_e = \frac{2}{3}\varepsilon_e = \frac{2}{3}N_e a' k_BT/V = \frac{7}{4}\frac{1}{3}\varepsilon_{\mathrm{ph}} \propto 1/R^4 \\ &\text{If } m_ec^2 \gg a'k_BT, \quad \text{then } p_e \approx \frac{3}{2}N_\ell k_BT, \quad \varepsilon_e \approx \varepsilon_\ell \approx N_\ell m_ec^2/V \propto 1/R^3 \end{aligned} \tag{13.8}$$

Thus, for T much higher than $m_ec^2 \approx 5.93 \times 10^9$ K the first line of (13.8) is valid, while for T much lower than 5.93×10^9 K the second line of (13.8) is valid where N_ℓ is the electron lepton number. We repeat here that observational data show that the number of photons is about a billion times larger than the number of baryons ($N_{ph} \approx 1.65 \times 10^9 N_B$). It follows from (13.8) that the electron/positron contribution is equal to 7/4 of the photon contribution for $T \gg 5.93 \times 10^9$ K and it is negligible for $T \ll 5.93 \times 10^9$ K.

For the unknown particles of the heavy dark matter it is not possible to make reliable predictions. Nevertheless, we shall assume that a behavior of the form $\varepsilon \propto 1/R^3$ is valid since much earlier times as a result of the absence of E/M and strong interactions. If the unknown particles of dark mass are of the supersymmetric type, their mass must be much larger than m_B. Anyway, there is strong evidence that the dark energy contribution to the energy density is about five times the baryonic contribution at least for $T \approx 10^{12}$ K or less. These assumptions are consistent with observational data. Based on the above relations and the corresponding data we find that the baryonic plus heavy dark matter contribution to the energy density becomes again comparable to the photonic one for $T \approx 7400$ K.

Finally, for the light dark matter, the neutrinos and antineutrinos, a formula similar to (13.8) is valid, but with two important quantitative differences: (a) The neutrino rest energy is extremely small, possibly less than 0.1 eV. (b) The number of neutrinos are comparable to the number of photons, as we argue before. Thus the contribution of the neutrinos to the energy density is comparable to that of photons and of the form $1/R^4$ down to low temperatures (possibly 1,000 K or lower). Below this threshold temperature their contribution is of the form $\varepsilon_\nu \approx N_\nu m_\nu c^2/V \propto 1/R^3$, since their temperature is not much different from that of photons.

The above remarks imply that the evolution of the Universe can be separated, after the inflation period, in three distinct epochs according to which component of the energy density is controlling its evolution:

a. Photon (and various matter/antimatter contributions depending on the temperature) dominated period for $T \gg 7400\,\mathrm{K}$, $\varepsilon \propto 1/R^4$.
b. Matter (baryonic and heavy dark matter) dominated period for $4\boldsymbol{K} \ll T \ll 7400\boldsymbol{K}$, $\varepsilon \propto \boldsymbol{1}/R^3$. At some time during this period the neutrino contribution probably changed from $1/R^4$ to $1/R^3$.
c. Dark energy dominated period $T \ll 4\,K$, $\varepsilon \approx \varepsilon_{\text{de}} = \text{const.}$

d. For the first period, (13.1) becomes $R\dot{R} = \text{const.}$ or $d(R^2)/dt = \text{const.}$ or

$$R(t) = At^{1/2}, \quad T(t) = C/R(t) = (C/A)\,t^{-1/2}, \quad \varepsilon \propto t^{-2} \tag{13.9}$$

where the constant C is approximately equal to 1.186×10^{27} K · m. The coefficient A is roughly equal to $A \approx 4.735 \times 10^{20}\,\mathrm{m \cdot y^{-1/2}}$ or $A \approx 8.43 \times 10^{16}\,\mathrm{m \cdot s^{-1/2}}$. With these rough choices of the coefficients A and C the radius corresponding to the temperature of 7,400 K, at which the matter contribution to the energy density becomes about equal to the photonic plus neutrino contribution, is $R = 1.6 \times 10^{23}$ m and the time at which this happens can be estimated to be 114 kyr. Up to this time, it is not unreasonable to still use the relation $R \propto \sqrt{t}$ for a rough estimate.

e. Going now to the matter dominated period, extending at least from 0.7 Myr to 5 Gyr, (13.1) becomes $R^{1/2}\dot{R} = \text{const.}$ or $d(R^{3/2})/dt = \text{const.}$ or

$$R \propto t^{2/3}, \quad T \propto t^{-2/3}, \quad \varepsilon \propto t^{-2} \tag{13.10}$$

f. Finally, for the dark energy dominated period, extending beyond 20 Gyr, (13.1) becomes $\dot{R} = \text{const.}\,R$ or

$$R \propto \exp(Ht), \quad T \propto \exp(-Ht), \quad \varepsilon \approx \varepsilon_{\mathrm{de}} = \mathrm{const.} \tag{13.11}$$

where $H = \sqrt{b\varepsilon_{\mathrm{de}}}$ is the Hubble constant and $b \equiv 8\pi G/3c^2$; during this dark energy period the Hubble constant is constant not only in space but in time as well.

To have more accurate results for the time dependence of R, T, and ε we need to interpolate between the various periods: from the inflation period to the photon (and matter/antimatter) period, from the latter to the matter period and from the matter period to the dark energy period. This requires detailed numerical calculations, Nevertheless, as we will see, the matter and dark energy periods can be treated analytically in a joint way, if the neutrino contribution is ignored. Indeed, from (13.1), we have that $\dot{R}/R = (b\varepsilon)^{1/2}$, where $b\varepsilon = aR^{-3} + g$, $a \equiv 6b(3/4\pi)N_B m_B c^2$, $g \equiv b\varepsilon_{\mathrm{de}}$. The factor 6 in a is there because we have included the heavy dark energy density which is taken to be five times the baryon energy density $N_B m_B c^2/V$. Thus, with this convenient notation, (13.1) takes the form,

$$\dot{R} = \left[(a/R) + gR^2\right]^{1/2} \tag{13.12}$$

The reader may verify that (13.12) is satisfied by the following relation:

$$R(t) = (a/g)^{1/3}\left[\sinh\left(\frac{3}{2}\sqrt{g}t\right)\right]^{2/3} \tag{13.13}$$

Notice that for $\frac{3}{2}\sqrt{g}t \ll 1$, (13.13) reduces to $R(t) \approx \left(\frac{3}{2}\right)^{2/3} a^{1/3} t^{2/3}$ in agreement with (13.10). Similarly, for $\frac{3}{2}\sqrt{g}t \gg 1$, we have $R(t) \approx 2^{-2/3}(a/g)^{1/3}\exp\left(\sqrt{g}\,t\right)$, in agreement with (13.11).

We can obtain explicit numerical results by substituting in (13.13) the best estimates of the present values for a/g and for $\sqrt{g}$. We have from $b\varepsilon = aR^{-3} + g$ that $(a/g)^{1/3} = R(\varepsilon_m/\varepsilon_{\mathrm{de}})^{1/3} = R_{\mathrm{pr}}(\varepsilon_m/\varepsilon_{\mathrm{de}})^{1/3}_{\mathrm{pr}} \approx R_{\mathrm{pr}}(0.279/0.721)^{1/3} = 0.729R_{\mathrm{pr}}$, where $\varepsilon_{m,\mathrm{pr}} \approx \varepsilon'$. Taking into account that presently $\varepsilon_{\mathrm{de}} \approx 0.721\varepsilon$ and that $H_{\mathrm{pr}} = \sqrt{b\varepsilon} \approx \sqrt{b\varepsilon_{\mathrm{de}}}/\sqrt{0.721}$ we have for $\frac{3}{2}\sqrt{g} \equiv \frac{3}{2}\sqrt{b\varepsilon_{\mathrm{de}}} = \frac{3}{2}(0.721)^{1/2}H_{\mathrm{pr}} \approx 3.057 \times 10^{-18}s^{-1} = 0.0964\,\mathrm{Gyr}^{-1}$. Thus (13.13) becomes

$$R(t) \approx 0.729R_{\mathrm{pr}}[\sinh(0.0964t/1\,\mathrm{Gyr})]^{2/3} \tag{13.14}$$

Let us subject our formula (13.14) to a few checks: If we substitute for t the estimated present age of the Universe, $t = 13.7$ Gyr we find $R_{pr} \approx 1.16\,R_{pr}$, a 16% discrepancy. The ratio of the present radius over the radius at $t = 0.38$ Myr is 1,240, according to (13.14), while the accepted value is 1,089, a 13% error. Notice that (13.14) is quite inaccurate for 'short' times such as 0.38 Myr, since detailed calculations indicate that at $t = 0.38$ Myr the neutrino and the photon contributions to ε are not negligible but they are 10 and 15%, respectively. The time t_e at which $\varepsilon' \approx \varepsilon_{\mathrm{de}}$ is obtained by noticing that the equality $\varepsilon' \approx \varepsilon_{\mathrm{de}}$ implies that $g \approx a/R^3$ or $R \approx (a/g)^{1/3}$ which in combination with (13.13) leads to $\sinh\left(\frac{3}{2}\sqrt{g}t_e\right) \approx 1$ or $t_e \approx 9.14\,\mathrm{Gyr}$, which is in rough agreement with more accurate calculations.

Finally, let us find the time t_o at which the acceleration vanishes, so that for $t < t_o$ the acceleration is negative, while for $t > t_o$ is positive. Equation (13.7) shows that $\ddot{R} = 0$ is equivalent to $\varepsilon + 3p = 0$, or $\varepsilon' + \varepsilon_{\text{de}} - 3\varepsilon_{\text{de}} = 0$, or $\varepsilon' = 2\varepsilon_{\text{de}}$. To obtain these relations we have taken into account that for t equal to a few billion years the only appreciable contribution to the pressure is the one coming from the dark energy. The relation $\varepsilon' = 2\varepsilon_{\text{de}}$ is equivalent to $a/R^3 \approx 2g$ or $R \approx (a/2g)^{1/3}$. Combining this last relation with (13.13) we obtain that $\sinh\left(\frac{3}{2}\sqrt{g}t_o\right) \approx 1/\sqrt{2}$ from which we get $t_o \approx 6.83\,\text{Gyr}$. This estimate is somehow lower than the best estimates for t_o.

13.4 Formation of the Structures of the Universe

Having the temperature and the density of the Universe as a function of time, we can employ known laws of physics to examine the formation of various structures of matter as time goes on. For the very early period of the Universe, $t \leq 10^{-14}$s, when the energy per particle was more than 10 TeV, rather detailed scenarios have been proposed including the inflation model, which was proven very effective in explaining several puzzling observations. These scenarios are based on theoretical schemes for which there is no direct experimental support. Here we shall restrict ourselves to briefly mention some crucial events in the history of the Universe starting from about 10^{-12} s corresponding to energies achievable experimentally.

Already in the big accelerators on Earth energies approaching the 10 TeV per particle have been achieved and experiments are under way where naked heavy nuclei moving with almost the velocity of light are forced to collide head-on with each other. Just after such a collision, the matter is in a state similar to that prevailing in the Universe during the period of about 10^{-12} to 10^{-6} s when the energy per particle was in the range from about 1.2 TeV to 1.2 GeV. For such energies quarks and gluons form a dense and very hot "soup" called *quark-gluon plasma*.

As the energy per particle dropped in the range 1 to 0.1 Gev corresponding to time in the range 10^{-6} to 10^{-4} s the process of *baryogenesis* started, during which quarks were combined, with the intervention of gluons, to form protons, neutrons and other short-lived baryons. During baryogenesis a thermodynamic equilibrium between protons and neutrons was established through reactions of the form $p + e^- \rightleftarrows n + \nu$ which lead to a ratio of neutrons to protons equal to $n/p = \exp(-\Delta\varepsilon/k_B T)$, where $\Delta\varepsilon = (m_n - m_p)c^2 = 1.2934\,\text{MeV}$, is the rest energy difference between neutron and proton. As the temperature was dropping this ratio was changing; at $t \approx 3s$, when $k_B T \approx 0.7\,\text{MeV}$, $n/p \approx 1/6$. From this point on the rate of the expansion of the Universe was faster than the rate of the reactions changing neutrons to protons and vice versa; thus the ratio n/p "froze" temporarily at the value 1/6. Then the neutrons start decaying irreversibly through the reaction $n \to p + e^- + \bar{\nu}$ until the nucleosynthesis started and the neutrons were trapped and stabilized within the He^4 nucleus, at about $t \approx 150$ s;at this point $n/p \approx 1/7$.

The *nucleosynthesis* started at about 120 s when $k_B T \approx 0.11$ MeV was low enough for the process to start with the reaction $n + p \rightarrow d + \gamma$; then it proceeded through a series of reactions which were concluded at about 1,200 s with almost all the neutrons bound in He^4 and in traces of deuterium. Thus, out of two neutrons and 14 protons (ratio 1/7) two neutrons and two protons made one He^4 nucleus and 12 protons remained free; the result is one He^4 nucleus for each dozen of protons or a mass ratio of 25 to 75% between He^4 and hydrogen, a prediction consistent with observational data.

As the Universe was expanding and was cooling electrons started to be bound around the He^4 nuclei and around the protons to form *neutral atoms* of helium and hydrogen. This process was concluded when the temperature was about 3,000 K; at this temperature one can show that thermodynamic equilibrium dictates that practically all matter consists of neutral atoms of He and hydrogen. The temperature of 3,000 K was reached when the radius of the observable Universe was about 1,090 smaller than the present radius and the time was about $t = 380{,}000$ yr. ((13.14) would give $t = 460{,}000$ yr for $R_{\text{pr}}/R \approx 1090$). The radius of the observable Universe at $t = 380{,}000$ yr was $R = 4 \times 10^{23}$ m and the concentration of neutral hydrogen atoms was 2.71×10^8 m^{-3} i.e. 1090^3 times larger than the present one which is about 0.21 m^{-3}.

For $T = 3{,}000$ K the maximum of the spectral distribution of the CBR appears at $\lambda \approx 9650$ A. From (5.16) of Chap. 5 we can find that the total scattering cross-section of photons of wavelength 9,650 A by neutral hydrogen atoms is equal to 6.58×10^{-33} m^2. The scattering cross-section by He^4 neutral atoms is 9% of that of hydrogen. The mean free path of photons at $t = 380{,}000$ yr can be calculated from (5.18) and it is equal to $\ell \approx 5.6 \times 10^{23}$ m, i.e., slightly larger than the radius of the observable Universe at that time. The conclusion is that at $t \approx 380000$ yr the Universe became transparent to E/M radiation or, equivalently, matter and light were decoupled. If, at this time, the matter were fully ionized the scattering by electrons would dominate, as shown by (5.9), and the mean free path would be 4.8×10^{19}, i.e., four orders of magnitude smaller than the radius of the observable Universe. Thus the matter by becoming neutral practically ceased to produce any light and, hence, became invisible. The dark age of the Universe started.

This dark age lasted for about 100 Myr or so, according to the latest estimates, until the first protostar was formed and reionized the matter around and inside it. The formation of galaxies as well as stars within the galaxies followed; stars during their lifetime fuse hydrogen and He to heavier elements all the way to iron. Beyond iron the fusion of heavier elements requires external supply of energy as shown in Fig. 7.1. This extra energy is provided during the explosive death of big stars through the supernova II process, which is credited for the creation of the heavier than iron nuclei and atoms. The material dispersed by the supernova explosions—containing all the chemical elements—is recycled by gravity to form new stars and planets such as ours acting as possible cradles of life and eventually of intelligent life wondering about this World, this small world, the great.

Epilogue: The Anthropic Principle

This World, he said, is your commandment and in your bowels is written.

O. Elytis

Abstract Do the Laws of physics and of the values of the Universal physical constants fit the phenomenon of life?

Summary The biological fitness of the laws of Nature and the biological fitness of the values of the universal physical constants constitute a part of one version of the anthropic principle. A stronger version claims that the statement 'Life exists' is not only the shortest but the most comprehensive statement about the World. Most versions of the anthropic principle include also an 'explanation' of either the fitness or the assumed necessity of the physical laws and the values of the physical constants for the phenomenon of life as we know it on Earth.

In our short journey over the structures of the World we bypassed the most interesting ones, namely the living structures. The reason was that living structures do not seem to fit with what was the purpose of this booklet: *To derive some select but important quantitative (or semi-quantitative) results regarding structures of matter in a simple way starting from a few established general laws of Nature and the values of the universal physical constants.* The out of thermodynamic equilibrium living matter seems to be too complex and diverse to fit within this limited purpose and thus allow a simple derivation of quantitative results from a few fundamental laws of physics.

However, the opposite direction, from the fact that "life exists" to the "laws of nature and to the values of the physical constants" is not without some merit [1.1, E1]. It must be admitted that this subject known under the generic name **anthropic principle** is rather controversial, because, on the one hand it is ideologically loaded, and on the other, it does not usually provide concrete quantitative conclusions. Nevertheless, in this short epilogue we shall mention a few points showing that one can deduce some fundamental laws of Nature or at least demonstrate their necessity from the fact that life exists. In what follows we shall refer mostly to the system Earth/Sun where carbon based life exists. However,

E. N. Economou, *A Short Journey from Quarks to the Universe*, SpringerBriefs in Physics, 1, DOI: 10.1007/978-3-642-20089-2,

most of our arguments are applicable to any other planet/star or satellite/star system sustaining earth-like life.

The admirably precise reproducibility from generation to generation of living structures strongly suggests that matter is "digital" not "analog", meaning that matter is of discrete not of continuous character. It is very difficult to imagine a continuous complex medium in contact with an ever changing environment to emerge again and again with exactly the same form and the same features. For this impressive phenomenon to occur, the huge amount of information stored in the "central bank" of the DNA molecule has to be kept intact. This in turn requires the double discretization shown in Fig. 3.1 (d or d′) because, otherwise, the DNA molecule would undergo a continuous change and it would never be the same. In other words, from the precise reproducibility of the living structures one is led to the discretization along both the vertical and the horizontal axis of Fig. 3.1(d or d′) The discretization along the vertical axis is consistent with the atomic idea. Similarly, the discretization along the horizontal axis is consistent with the wave-particle duality [E1]. Whether one can reverse the argument and deduce from the double discretization the atomic structure of matter (and forces) as well as the particle/wave duality is not clear.

The existence of an extremely weak, attractive, long range force, such as gravity, seems to be necessary in order to keep the atmosphere, the lakes, the oceans, and the living matter on the surface of the Earth; as well as to keep the Earth going around the Sun and drawing from it both the energy and the information required for the emergence and maintenance of life. This leads to the necessity of another force, the electromagnetic one, which has a dual responsibility: To transmit the energy and the information from the Sun to the Earth; and to bring and keep together all matter including the one from which the living structures consist. The strength of the E/M force has to be enormous in comparison with gravitational one. This becomes apparent by considering what the size of a single-cell organism would be, if it would be held together by gravitational forces instead of the electromagnetic ones: its diameter would be an absurd size of 10^{21} m, instead of 10^{-5} m which is its actual size! Moreover, the E/M force by being both attractive and repulsive manages to be both long range in transmitting energy and information from the Sun and short range in bringing and keeping together the constituents of living matter.

Life on Earth requires the continuous flow of energy from the Sun over a long period (of the order of a few billion years). Such a supply of energy cannot be chemical, because it would be exhausted in a period of thousands of years. There must be another source of energy about a million times larger than the chemical one. Assuming that the available energy is of the same order of magnitude as the kinetic energy and taking into account the Heisenberg principle, we conclude that the required supply of energy E per particle must come from a region of size r and from particles of mass m such that $E/E_{chem} \approx \left(m_e a_B^2/m\, r^2\right) \approx 10^6$. The nuclear size of $r = 1$ fm and the proton mass of 1836 m_e produces the necessary boost of 1.5×10^6. Hence, it is clear that the nucleus is the source of the energy supply in the Sun (and in any other active star). This implies that nuclei are composite particles held together by a third type of force- the strong one capable to overcome the kinetic energy of the nucleons and the Coulomb repulsion among the protons. Finally, a fourth interaction, the weak one, is necessary in order to allow the transformation from proton to neutron and vice versa. Without this, the fusion of two protons to a deuteron (with the emission of a positron and a neutrino) would not be possible, neither would be possible the eventual fusion of four protons to a helium nuclei, the main reaction providing the energy of the stars.

Let us consider now the values of some universal physical constants. The mass of the electron is less than the mass difference between a neutron and a proton. This allows the neutron, if free, to decay to a proton, an electron, and an antineutrino. If this were not the case, electrons and protons would combine to free neutrons in the early Universe leaving very few protons to act as fuel in the stars. Thus, a neutron has to be heavier than the sum of an electron and a proton but not much heavier, because in such a case very few neutrons would be produced during the baryogenesis or within the nuclei [see (7.2)]. As a result of the greater percentage of protons heavier elements necessary for life would not be stable. Elements heavier than He and up to Fe are made by fusion within active stars. This stellar nucleosynthesis passes through a bottle-neck to reach the carbon nucleus. The reason is that the Be^8 nucleus consisting of four protons and four neutrons is extremely metastable with a life-time of the order of 10^{-15} s, because it is energetically favorable to break into two He nuclei. Thus it is extremely improbable for a third He atom to be incorporated in the Be^8 to produce the crucial for life carbon atom, unless a resonance effect is present. Indeed, such an effect occurs in the sense that the energy of Be^8 plus the energy of the third He nucleus coincides with an excited energy level of carbon at 7.65 MeV above its ground state. This coincidence greatly enhances the probability of the incorporation of the third He nucleus and makes the carbon nucleus possible. It is worth mentioning that Fred Hoyle *predicted* the occurrence of such an excited level in the carbon nucleus before it was found experimentally.

The existence of life implies that the dimensionless gravitational constant must have a value close to the measured one. Indeed, it has been estimated that about 10 billion years are needed for life to emerge in the Universe, since life is the final stage and the crowning of the matter dominated epoch. On the other hand, life ought to appear and evolve before the dark energy dominates, since the latter tends to disolve the structures of matter. These two requirements imply that $\sqrt{g}\,t \approx 1$ for $t \approx 10^{10}$ ys [see (13.13)]. This in turn means that $a_G \approx 4.7 \times 10^{-39}$ i.e. about 20% smaller than its measured value. (The estimated value of $\rho_{de} = \varepsilon_{de}/c^2 \approx 0.721 \times 1.034 \times 10^{-26}\mathrm{kg/m^3}$ was used in obtaining this value of a_G).

As it was mentioned before, supernova explosions, in particular core-collapse type II supernova explosions (see below), are important for the creation and the recycling of the elements in general and for the emergence of life in particular. The heavier than iron elements are created during such explosions. Moreover, all elements are dispersed as a result of supernovae very efficiently in the interstellar medium to be recycled for the production of the next generation of stars, of planetary systems and, eventually, of humans. In rough terms, a Type II supernova explosion occurs as follows: In a big star upon the exhaustion of its nuclear fuel, its central part collapses to a core so dense that even neutrinos are trapped in it. This superdense core attracts and accelerates the outer part of the star, which hits the core at high speed. This internal collision forces the outer part to rebound and at the same time releases the neutrinos which push the outer part with explosive force leading to the supernova II. For the neutrino flux to be capable of catapulting the outer part of the star against the gravitational attraction, the following relation connecting the weak dimensionless coupling constant a_W with the gravitational one a_G, must be satisfied:

$$a_W^4 \approx a_G \left(m_p/m_e\right)^6 \quad \Rightarrow \quad a_W \approx 2.18 \times 10^{-5}$$

The resulting value of a_W is in reasonable agreement with its theoretical and measured value.

We shall conclude this epilogue by drawing attention to the mystery of the cosmological constant Λ, or equivalently the density of the dark energy, ε_{de}. If we accept the established value of the gravitational constant, G, the measured value of Λ has the right order of magnitude for the transition from matter domination to dark energy domination to occur at $t \approx 10$ billion years, i.e., when the time was ripe for life to emerge. This looks like an extremely fine tuning of the value of Λ, in view of theoretical considerations producing values of Λ 10^{120} larger than the actual one!

Appendix I: Oscillations and Waves

Oscillations occur locally in systems where their energy can change back and forth from the form of potential energy to the form of kinetic energy. A typical mechanical example is the pendulum shown in Fig. I (a). A typical electrical example is the *LC* circuit shown in Fig. I (b); in this case the potential energy is equal to $\frac{1}{2}Q^2/C$, while the kinetic-type energy is $\frac{1}{2}LI^2/c^2$ (which includes both the systematic kinetic energy $\frac{1}{2}N_e m_e \upsilon^2$ of the electrons and the magnetic energy associated with the current $I = (N_e/\ell)\, e\, \upsilon$). C is the capacitance, L is the self-inductance $I = dQ/dt$, and ℓ is the overall length of the wire making up the circuit. In cases of macroscopic oscillations, as those in Fig. I(a) and (b) macroscopic kinetic energy is transferred through collisions with microscopic particles to the internal energy of both the system and the environment. This process is described macroscopically as the action of frictional "forces", which eventually lead to the "loss" of the macroscopic oscillating energy and the eventual termination of the macroscopic oscillation. If there is no transfer of energy from the oscillating system, then conservation of energy, $E_t = E_K + E_P$ implies that $E_{KM} - E_{Km} = E_{PM} - E_{Pm}$, where the subscripts M and m indicate maximum and minimum, respectively. This observation allows us to obtain the angular frequency $\omega = 2\pi f$ (f is the frequency) of the oscillation. E.g., in the case of Fig. I(b) by setting $LI_M^2/c^2 = Q_M^2/C$, and taking into account that $I^2 = \omega^2 Q^2$, we obtain $\omega^2 = c^2/LC$. (See also Sect. 5.4, p. 39.)

Waves are oscillations which migrate from local subsystems to neighboring subsystems within an extended medium (and, consequently, spread and are delocalized). The medium can be discrete, as the coupled pendulums shown in Fig. I(c), or continuous, as in the case of sea-waves shown schematically in Fig. I(d). A wave in addition to the angular frequency ω of the propagating oscillation is usually characterized by its wavelength λ (or equivalently by its wavenumber $k \equiv 2\pi/\lambda$) and its velocity υ of propagation. The three quantities ω, k, υ are related as follows:

$$\omega = \upsilon k \tag{I.1}$$

E. N. Economou, *A Short Journey from Quarks to the Universe*,
SpringerBriefs in Physics, 1, DOI: 10.1007/978-3-642-20089-2,

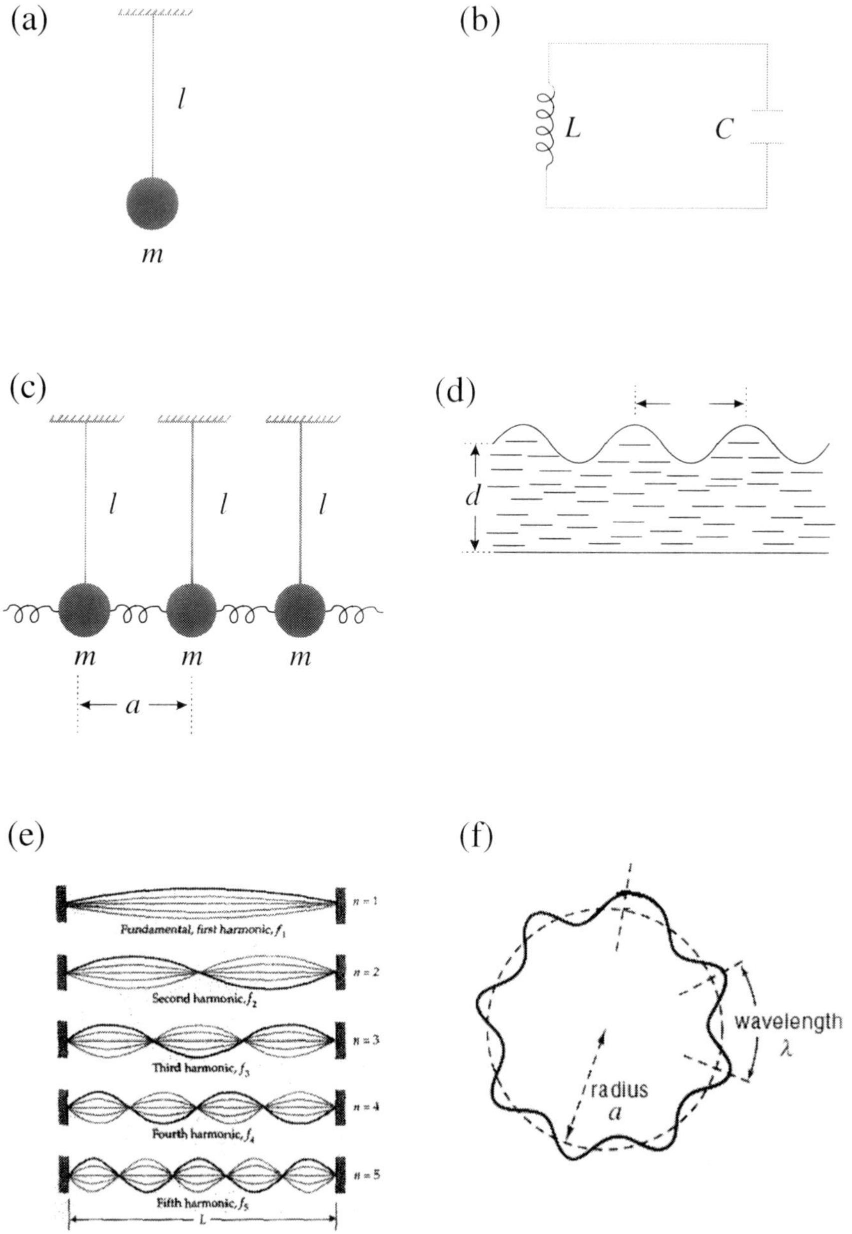

Fig. I **a** A pendulum. **b** An LC circuit. **c** A medium of coupled pendulums. **d** Sea waves. **e** Standing or stationary waves in a string of fixed ends. **f** Standing or stationary wave around a circle

The velocity υ depends on the properties of the medium and possibly of the wavenumber k. (See also Sect. 5.4, p. 39.)

Waves confined in a finite region of space have to satisfy certain boundary conditions, which restrict the allowed values of the angular frequency to a discrete set, ω_o, ω_1, ω_2, ... (see Fig. 3.1, p. 23). To each value ω_i correspond one or more waves, called stationary (or standing) waves which, having spread over the available space, are not traveling anymore. Examples of standing waves are the ones appearing in a string of fixed ends [see Fig. I(e)] on the ones going around in a circle [see Fig. I(f)].

Selected Problems

4.1 Taking into account (4.12), plot schematically the Gibbs free energy as a function of temperature under constant pressure for the three phases of matter. Then, obtain the phase diagram in the plane T, P.
4.2 Prove that $\Omega = -PV$.
5.1 Estimate the drag force on an object moving with velocity υ within a fluid.
5.2 Obtain a formula for the viscosity of a liquid. Dimensions of viscosity are pressure$\times$ time. Estimate the viscosity of water at $T \approx 295\,K$.
5.3 Estimate the surface-tension coefficient of water. See Sect 7.1.
5.4 Obtain the so-called Poiseuille's formula giving the volume of running fluid passing per second through any cross-section of a pipe.
5.5 High frequency current density flowing in a metallic wire of conductivity σ is confined in a region near the surface of depth δ. Obtain δ, called skin-depth.
5.6 Estimate the value λ_m at which the maximum of the black body radiation plotted as a function of the wavelength appears. Similarly, for ω_m in the I_ω versus ω curve. Is $\omega_m = 2\pi/\lambda_m$? If not, why?
5.7 Estimate the life-time of a classical model of a hydrogen atom. If there were no radiation, the electron will follow a circular orbit of radius a_B.
5.8 Estimate the lifting force on the wings of a plane. Assume that the wings are rectangular flat rigid metallic sheets.
6.1 Consider the baryons Δ^{++}, Σ^o, Ξ^-. What is their composition? What is the Feynman diagram describing their decay? Can you estimate their life-time?
6.2 The same for the mesons π^o, π^+, π^-, K^o, $\bar{K}^o$, K^+, J/ψ.
7.1 The composition of natural uranium is 99.3% U-238 and 0.7% U-235. Their half-lifes are 4.51×10^9s and 7.1×10^8 s respectively. Obtain limits for the age of our planetary system and the age of the Universe.
7.2 The distribution of the mass number of the fragments of the neutron induced fission of U-235 in a nuclear reactor exhibits a double peak at $A = 92$ and $A = 140$. (Eisberg-Resnick, *Quantum physics*.) On the contrary, the fragments of a fission bomb exhibit a single peak at $A \approx 116$. Any explanation?

E. N. Economou, *A Short Journey from Quarks to the Universe*,
SpringerBriefs in Physics, 1, DOI: 10.1007/978-3-642-20089-2,

8.1 Estimate the first, the second, and the third ionization potential for the atom of Li. (See Fig. 8.2).

9.1 The energy E vs. distance d between two argon atoms is $E = 4\varepsilon\left[(\sigma/d)^{12} - (\sigma/d)^{6}\right]$, where $\varepsilon = 10.4\,\text{meV}$ and $\sigma = 3.4\,\text{A}$. Calculate (a) the bond length (b) the oscillation frequency (c) the dissociation energy (d) the fluctuation of the bond length. Will the molecule survive at room temperature?

9.2 Identify the four eigenoscillations of CO_2. Classify them according to the size of their frequency. Which one is responsible for the greenhouse effect?

9.3 Determine the stereochemistry of C_2H_2, C_2H_4, C_2H_6, C_4H_6, C_4H_{10}.

9.4 For the NaCl molecule the relevant atomic levels and the bond length are, $\varepsilon_{\text{Na}} \approx -4.95\,\text{eV}$, $\varepsilon_{\text{Cl}} \approx -13.78\,\text{eV}$, $d = 2.36\,\text{A}$. Determine the bonding and antibonding molecular levels, the polarity index and the charge transfer from Na to Cl. The ionization potential of Na is 5.14 eV and the chemical affinity of Cl is 3.61 eV. What is the dissociation energy of NaCl?

9.5 In analogy with (9.8) the equations satisfied by the 6 p_z orbitals in benzene are $(\varepsilon - \varepsilon_\nu)c_n^{(\nu)} + V_2 c_{n-1}^{(\nu)} + V_2 c_{n+1}^{(\nu)} = 0$, $n = 1, \ldots, 6$; $c_7^{(\nu)} \equiv c_1^{(\nu)}$. Show that Eqs. 9.15 satisfy these equations with $\varepsilon_\nu = \varepsilon_p + 2V_{2,zz}\cos\varphi_\nu$. Draw the levels ε_ν for the 6 values of ν.

10.1 The low-temperature specific heat of a metal is of the form $C = a_1 T + a_2 T^3$, where the first term is due to electronic excitations and the second to vibrational ones. Using the principles of Pauli and Schrödinger explain this behavior.

10.2 Make a rough estimate of the melting temperature of a solid.

10.3 In a diagram show the relevant atomic levels, the hybrid level, the bonding and antibonding molecular levels of Si, as well as its valence and conduction bands. What is the value of the energy gap in Si?

10.4 The same for the III–V semiconductor GaAs. (see Table 4; $d = 2.45$ A).

10.5 What is the value of $\bar{r}$ for liquid water? What is the velocity of sound in water (keep in mind that the hydrogen bond is 3 to 4 times weaker than a typical metallic bond)? What is the value of the bulk modulus of water?

10.6 Consider pure crystalline Si. Replace one Si atom by a P atom. Four of the valence electrons of P will participate in the four tetrahedral bonds. The fifth one will be detached from the immediate vicinity of the parent atom, but will be trapped around it at a distance r much larger than a_B. Taking into account that the dielectric constant of Si is 13.1 and that the effective electron mass is $0.32m_e$, find the distance r and the corresponding binding energy. If one atom of Si in every million is replaced by a P atom, what would be the effect on the resistivity at room temperature?

11.6 Mars's year is longer than Earth's by a factor of 1.88. What is the temperature at the surface of Mars?

11.7 The period of Moon around the Earth is 27.32 days, while the period of full moon is 29.5 days. Provide a quantitative explanation for this difference.

11.8 The period of tides is about 12 hours. Why? When is the phenomenon of tide stronger? At full-moon? At new-moon? At half-moon? Explain.

12.1 What would be the radius and the entropy of Earth, if it would become a black hole?

12.2 For an active star in the main sequence (see [11.2] or [12.1]) we have the following approximate formulae: Pressure at their center, $P_c \approx 2\rho GM/R$. Temperature at their center, $T_c \approx P_c m/k_B \rho$. Radius, $R \propto M^a$; $a \approx \frac{1}{2}$. Surface temperature, $T_S \propto M^\beta$; $\beta \approx \frac{1}{2}$. Luminosity, $L \propto R^2 T_S^4 \propto M^{2a+4\beta}$; $2a + 4\beta \approx 4$ for $0.5M_S \leq M \leq 2M_S$ and 3.5 for $2M_S \leq M \leq 20M_S$. Life-time $t \approx M/L$. Can you justify these formulae. Apply them to our Sun. Do they work?.

12.1 What was the pressure of baryons, of photons, and of neutrinos at $t \approx 380\,\text{kyr}$?

13.2 Consider a photon gas in thermodynamic equilibrium with electron/positron pairs. The temperature is much higher than $m_e c^2/k_B$ and the lepton number is negligible. The energy of each species is: $E_{ph} = 2\sum_k \varepsilon_k b_k$, $E_e \approx E_{e^+} = 2\sum_k \varepsilon_k$, $f_k \varepsilon_k = \hbar c k$, $b_k = [\exp(\varepsilon_k/k_B T) - 1]^{-1}$, $f_k = [\exp(\varepsilon_k/k_B T) + 1]^{-1}$. Justify the above formulae. Taking into account that $\sum_k \to (V/2\pi)^3 \int d^3k$ show that $E_e = \frac{7}{8} E_{ph}$.

Bibliography

P1 S.W. Hawking, L. Mlodinow, *The grand Design* (Bantam books, NY, 2010)
1.1 P. Morrison, P. Morrison, *Powers of Ten* (Scientific American Books, NY, 1982)
2.1 M. Veltman, *Facts and Mysteries in Elementary particle physics* (World Scientific, New Jersey, 2003)
3.1 R. Feynman, R. Leighton, M. Sands, *The Feynman Lectures on Physics* (Addison-Wesley, Reading, 1964)
3.2 P. Ghose, *Testing Quantum Mechanics on New Ground* (Gambridge University Press, Cambridge, 1999)
3.3 R. Griffiths, *Consistent Quantum Theory* (Gambridge University Press, Cambridge, 2002)
3.4 H. Paul, *Introduction to Quantum Theory* (Gambridge University Press, Cambridge, 2008)
3.5 A. Peres, *Quantum Theory: Concepts and Methods* (Kluwer Academic Publishers, Dordrecht, 1995)
3.6 A. Einstein, B. Podolsky, N. Rosen, Phys. Rev. **47**, 777 (1935)
3.7 A. Aspect, P. Grangier, G. Roger, Phys. Rev. Lett. **49**, (1982)
3.8 R. Shankar, *Principles of Quantum Mechanics*, 2nd edn. (Springer, NY, 1994)
4.1 E. Fermi, *Thermodynamics* (Dover Publications, NY, 1956)
4.2 L. Landau, E. Lifshitz, *Statistical Physics, Part 1* (Pergamon Press, Oxford, 1980)
6.1 D. Perkins, *Introduction to High Energy Physics* (Addison-Wesley, Reading, 1987)
6.2 W. Cottingham, D. Greenwood, *An Introduction to the Standard Model*, 2nd edn. (Cambridge University Press, Cambridge, 2007)
6.3 F. Wilczek, *The Lightness of Being: Mass, Ether, and the Unification of Forces* (Basic Books, New York, 2008)
7.1 R. Leighton, *Principles of Modern Physics* (McGraw-Hill, NY, 1959)
7.2 A. Das, T. Ferbel, Introduction to nuclear and particle physics
9.1 J. McMurry, *Organic Chemistry* (Brooks/Cole, Belmont, MA, 1996)
9.2 W. Harrison, *Electronic structure and the poperties of solids* (Dover Publications, NY, 1989)
10.1 E. Economou, *The physics of solids, essentials and beyond* (Springer-Verlag, Heidelberg, 2010)
11.1 E. Cohen, D. Lide, G. Trigg (eds.), *AIP physics desk reference* (Springer-Verlag, NY, 2003)
11.2 F. Shu, *The physical universe: an introduction to astronomy* (University Science Books, Mill Valley, CA, 1982)
12.1 B. Caroll, D. Ostlie, *An introduction to modern astrophysics* (Addison-Wesley, Reading, 2007)
12.2 S. Lloyd, Y.J. Ng, Scientific American, Nov. 2004, p. 53; J. D. Bekenstein, Scientific American, **289**, 58, Aug. 2003

E. N. Economou, *A Short Journey from Quarks to the Universe*,
SpringerBriefs in Physics, 1, DOI: 10.1007/978-3-642-20089-2,

13.1 S. Weinberg, *The first three minutes* (Fontana Paperbacks, London, 1983)
E1. E. Schrödinger, *What is life?* (Gambridge University Press, Cambridge, 1967)

The Castoriadis quote is from the Castoriadis and Evangelopoulos book, *Philosophy and Science* (Editions Eurasia, Athens, 2010)

Index

E. N. Economou, *A Short Journey from Quarks to the Universe*,
SpringerBriefs in Physics, 1, DOI: 10.1007/978-3-642-20089-2,

Table 1 Physical constants. Numbers in parentheses give the standard deviation in the last two digits

Quantity	Symbol	Value (year 2008)	Units
Planck constant over 2π	$\hbar = h/2\pi$	$1.054\,571\,628\,(53) \times 10^{-34}$	$J \cdot s$
	$\hbar$	$6.582\ 118\ 99(16) \times 10^{-16}$	$eV \cdot s$
Velocity of light	c	299 792 458	$m \cdot s^{-1}$
Gravitational constant	G	$6.674\ 28(67) \times 10^{-11}$	$m^3kg^{-1}s^{-2}$
Proton charge	e	$1.602\ 176\ 487(40) \times 10^{-19}$	C
Electron mass	m_e or m	$9.109\ 382\ 15(45) \times 10^{-31}$	kg
Proton mass	m_p	$1.672\ 621\ 637(83) \times 10^{-27}$	kg
Neutron mass	m_n	$1.674\ 927\ 211(84) \times 10^{-27}$	kg
Atomic mass constant $\frac{1}{12}m(C^{12})$	m_u(or u)	$1.660\ 538\ 782(83) \times 10^{-27}$	kg
Vacuum permittivity	e_0	$8.854\ 187\ 817\ldots \times 10^{-12}$	$F \cdot m^{-1}$
Vacuum permeability	μ_0	$4\pi \times 10^{-7}$	$N \cdot A^{-2}$
Boltzmann constant	k_B	$1.380\ 6504(24) \times 10^{-23}$	$J \cdot K^{-1}$
Avogadro constant	N_A	$6.022\ 141\ 79(30) \times 10^{23}$	mol^{-1}
Fine-structure constant	α^*	$(137.035\ 999\ 679(94))^{-1}$	
Magnetic flux quantum	$\Phi_0{}^*$	$2.067\ 833\ 667(52) \times 10^{-15}$	Wb
Quantum Hall resistance	$R_H{}^*$	25 812.8075(80)	Ω
Bohr magneton	$\mu_B{}^*$	$927.400\ 915(23) \times 10^{-26}$	$J \cdot T^{-1}$
Nuclear magneton	$\mu_N{}^*$	$5.050\,783\,24\,(13) \times 10^{-27}$	$J \cdot T^{-1}$
Electron magnetic moment	μ_e	$-1.001\ 159\ 652\ 181\ 11(74)$	μ_B
Proton magnetic moment	μ_p	2.792 847 356(23)	μ_N
Neutron magnetic moment	μ_n	$-1.913\ 042\ 73\ (45)$	μ_N
Gas constant	$R \equiv N_A k_B$	8.314 472(15)	$J \cdot mol^{-1} \cdot K^{-1}$
Bohr radius	$a_B \equiv 4\pi e_0 \hbar^2/m_e e^2$	$0.529\ 177\ 208\ 59(36) \times 10^{-10}$	m

$\alpha = e^2/4\pi e_0 \hbar c$, $\Phi_0 = h/2e$, $R_H \equiv h/e^2$, $\mu_B \equiv e\hbar/2m_e$, $\mu_N \equiv (m_e/m_p)\mu_B$ (all in SI)

$\alpha = e^2/\hbar c$, $\Phi_0 = hc/2e$, $a_B = \hbar^2/m_e e^2$, $\mu_B = e\hbar/2m_e c$, $\mu_N \equiv (m_e/m_p)\mu_B$ (all in G – CGS)

Table 2 Atomic system of units ($m_e = 1$; $e = 1$; $a_B = 1$; $k_B = 1$) $c = 1/\alpha$ in G—CGS; $\varepsilon_0 = 1/4\pi$, $\mu_0 = 4\pi\alpha^2$ in SI

Length	$l_0 = a_B$
Mass	$m_0 \equiv m_e$
Charge	$q_0 \equiv e$
Time	$t_0 \equiv m_e a_B^2/\hbar = 2.418\,884 \times 10^{-17}$ s
Energy	$E_0 \equiv \hbar^2/m_e a_B^2 = 4.359\,744 \times 10^{-18}$ J $= 27.211\,384$ eV
Angular frequency	$\omega_0 \equiv \hbar/m_e a_B^2 = 4.134\,137 \times 10^{16}$ rad/s
Velocity	$v_0 \equiv a_B/t_0 = \hbar/m_e a_B = \alpha c = 2\,187.691$ km/s
Mass density	$\rho_0 = m_e/a_B^3 = 6.147\,315$ kg/m^3
Temperature	$T_0 \equiv E_0/k_B = \hbar^2/m_e a_B^2 k_B = 315\,775$ K
Pressure	$P_0 \equiv E_0/a_B^3 = \hbar^2/m_e a_B^5 = 2.942\,101 \times 10^{13}$ N/m^2 $= 2.942\,101 \times 10^8$ bar
Electrical resistance	$R_0 \equiv \hbar/e^2 = R_H/2\pi = 4\,108.236\ \Omega$
Resistivity	$\rho_{\rho 0} = R_0 a_B = \hbar a_B/e^2 = 21.739\,848\ \mu\Omega \times$ cm
Conductivity	$\sigma_0 = 1/\rho_{c0} = e^2/\hbar a_B = 4.599\,848 \times 10^7\ \Omega^{-1}\text{m}^{-1}$
Electric current	$i_0 \equiv e/t_0 = 6.623\,618 \times 10^{-3}$ A
Voltage	$V_0 \equiv E_0/e = 27.211\,384$ V
Electric field	$\mathrm{E}_0 \equiv V_0/a_B = 5.142\,206 \times 10^{11}$ V/m
Magnetic field[1]	$B_0 \equiv c\hbar/e a_B^2 = 2.350\,517 \times 10^5$ T
Electric polarizability [2]	$a_{e0} = 4\pi e_0 a_B^3 = 1.648\,777 \times 10^{-41}$ Fm^2
Electric induction	$D_0 \equiv e/a_B^2 = 57.214\,762$ C/m^2
Magnetic moment	$\mu_0 \equiv 2\mu_B = 1.854\,802 \times 10^{-23}$ JT^{-1}
Magnetization	$M_0 \equiv \mu_0/a_B^3 = 1.251\,682 \times 10^8$ A/m
Magnetic field	$H_0 \equiv M_0 = 1.251\,682 \times 10^8$ A/m

For any quantity X we define $\bar{X} = X/X_0$ [see (2.16)]

[1] In SI set $c = 1$. [2] In G–CGS set $4\pi e_0 = 1$

Table 3 Periodic table of the elements

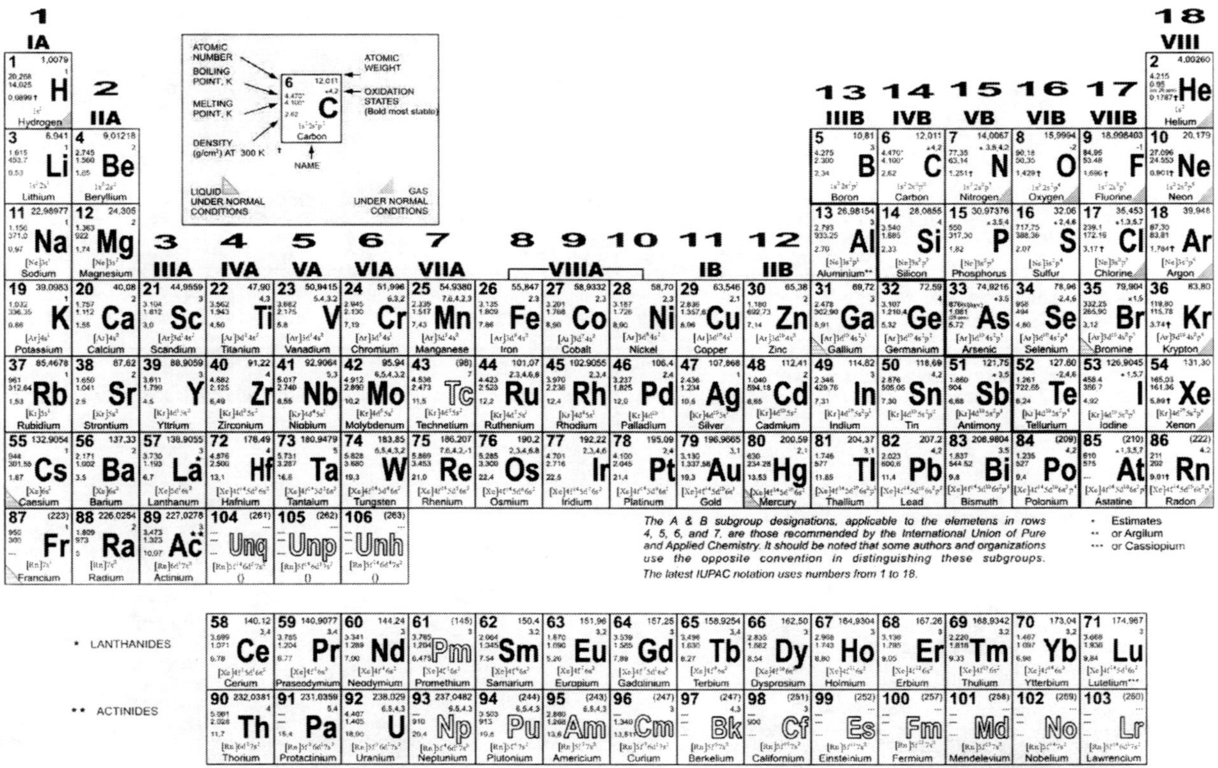

† For the gas phase (noted by the symbol ◢) the units is g/l. for T=273K and P=1atm.

Table 4 Theoretical values of the energy levels $-E_{n's}$, $-E_{n'p}$, $-E_{n'-1,d}$, and $V_1 \equiv (E_{n'p} - E_{n's})/4$ for the highest partially or fully occupied subshell (From Harrison' s book [20])

Non-transition elements (key):

C	
19.37	$\leftarrow -E_{n's}$
11.07	$\leftarrow -E_{n'p}$
2.08	$\leftarrow V_1$

Transition elements (key):

Fe	
7.08	$\leftarrow E_{n's}$
16.54	$\leftarrow E_{n'-1,d}$

H																	***He*** 24,97 -	
Li 5,34 - -	***Be*** 8,41 - -											***B*** 13,46 8,43 1,26	***C*** 19,37 11,07 2,08	***N*** 26,22 13,84 3,10	***O*** 34,02 16,72 4,33	***F*** 42,78 19,86 5,73	***Ne*** 52,51 23,13 7,35	$n' = 2$
Na 4,95 - -	***Mg*** 6,88 - -											***Al*** 10,70 5,71 1,25	***Si*** 14,79 7,58 1,80	***P*** 19,22 9,54 2,42	***S*** 24,01 11,60 3,10	***Cl*** 29,19 13,78 3,85	***Ar*** 34,75 16,08 4,67	$n' = 3$
K 4,01 - -	***Ca*** 5,32 - -	***Sc*** 5,72 9,35	***Ti*** 6,04 11,04	***V*** 6,32 12,55	***Cr*** 6,59 13,94	***Mn*** 6,84 15,27	***Fe*** 7,08 16,54	***Co*** 7,31 17,77	***Ni*** 7,52 18,96	***Cu*** 6,49 13,36	***Zn*** 7,96 4 0,99	***Ga*** 11,55 5,67 1,47	***Ge*** 15,15 7,33 1,96	***As*** 18,91 8,98 2,48	***Se*** 22,86 10,68 3,05	***Br*** 27,00 12,43 3,64	***Kr*** 31,37 14,26 4,28	$n' = 4$
Rb 3,75 - -	***Sr*** 4,85 - -	***Y*** 5,34 6,80	***Zr*** 5,68 8,46	***Nb*** 5,95 10,03	***Mo*** 6,19 11,56	***Tc*** 6,39 13,08	***Ru*** 6,58 14,59	***Rh*** 6,75 16,16	***Pd*** 6,91 17,66	***Ag*** 5,98 14,62	***Cd*** 7,21 4 0,803	***In*** 10,14 5,37 1,19	***Sn*** 13,04 6,76 1,57	***Sb*** 16,02 8,14 1,97	***Te*** 19,12 9,54 2,40	***I*** 22,34 10,97 2,84	***Xe*** 25,69 12,44 3,31	$n' = 5$
Cs 3,36 - -	***Ba*** 4,29 - -	***La*** 4,35 -	***Hf*** 5,72 8,14	***Ta*** 5,98 9,57	***W*** 6,19 10,96	***Re*** 6,38 12,35	***Os*** 6,52 13,73	***Ir*** 6,71 15,13	***Pt*** 6,85 16,55	***Au*** 6,01 14,17	***Hg(α)*** 7,10 4,11 0,75	***Tl*** 9,82 5,23 1,15	***Pb*** 12,48 6,53 1,49	***Bi*** 15,19 7,79 1,85	***Po*** 17,96 9,05 2,23	***At*** 20,82 10,33 2,62	***Rn*** 23,78 11,64 3,04	$n' = 6$
Fr	***Ra***	***Ac***																

Ce	***Pr***	***Nd***	***Pm***	***Sm***	***Eu***	***Gd***	***Tb***	***Dy***	***Ho***	***Er***	***Tm***	***Yb***	***Lu***
Th	***Pa***	***U***	***Np***	***Pu***	***Am***	***Cm***	***Bk***	***Cf***	***Es***	***Fm***	***Md***	***No***	***Lr***